WAY OUT WEST

IMAGES OF THE AMERICAN RANCH

WAY OUT WEST

CHARLIE SEEMANN

PHOTOGRAPHS FROM THE FARM SECURITY ADMINISTRATION
1936–1943

Guilford, Connecticut
Helena, Montana

A · TWODOT® · BOOK
An imprint of Globe Pequot
A registered trademark of Rowman & Littlefield

Distributed by NATIONAL BOOK NETWORK

British Library Cataloguing in Publication Information available

Library of Congress Cataloging-in-Publication Data available

ISBN 978-1-4930-2727-9 (hardcover)
ISBN 978-1-4930-2728-6 (e-book)

∞™ The paper used in this publication meets the minimum requirements of American National Standard for Information Sciences—Permanence of Paper for Printed Library Materials, ANSI/NISO Z39.48-1992.

Printed in the United States of America

ACKNOWLEDGMENTS

This book has been a number of years in the making, and there are several people that I need to thank for their inspiration, help, and support. First, I have to thank Larry and Maggie Biehl and the Hartman Foundation, without whose support this book would not have been possible. Second, my friend and mentor, the late Joe Wilson, who first encouraged me to work with FSA photographs and to create an exhibition of FSA photos of folklife in California back in 1979. Beverly Brannan and Erica Kelly at the Library of Congress were immensely helpful in acquiring copies of the photos. Thanks to Katherine Hearon for proofreading the manuscript. Finally, I want to thank my agent, Rita Rosenkranz, and my editor at Rowman & Littlefield, Erin Turner.

CONTENTS

HEADIN' FER THE NEW DEAL

(DEDICATED TO THE HON. FRANKLIN D. ROOSEVELT, PRESIDENT OF THE UNITED STATES)

We're headin' fer the NEW DEAL now. We've had some awful years.
We recollect them two cent cows, and them there four cent steers.
Besides the calves that got so cheap; the wust I've ever seen.
It made their mothers stand and weep the day that they was weaned.

We won't furgit them awful days; we're jest about all in.
We hope that they have gone to stay and won't come back agi'n.
Nobody seemed to onderstand. The folks was all in doubt.
There didn't seem to be no hand could git things straightened out.

It isn't always them that lags that gits a drive all sp'iled.
It ain't no use to charge the drags and let the lead run wild.
You hadn't ort to git ahead and hold 'em back too slow;
Fer then the herd takes on a spread and jest won't hardly go.

A new Trail Boss come ridin' in to soter see us through.
He see what we was up ag'in and figgered what to do.

Franklin D. knowed the herd was out of j'int. He steadied down the lead;
And put some old hands on the p'int to regalate the speed.

And then he put the flank and swing 'bout where they'd ort to ride.
Not too close up and crowdin' things, nor keepin' out too wide.
The drags is safe from fightin' stock, and crazy riders too;
They're goin' at a steady walk, they'll easy make it through.

He doesn't let 'em git TOO DRY. He keeps 'em on the feed,
He knows his herd and that is why he savvys what they need.
He saved us from a total loss. That's why the boys allow,
They're stringin' with the new Trail Boss that's in the saddle now.

BRUCE KISKADDON
COWBOY POET
DECEMBER 1933

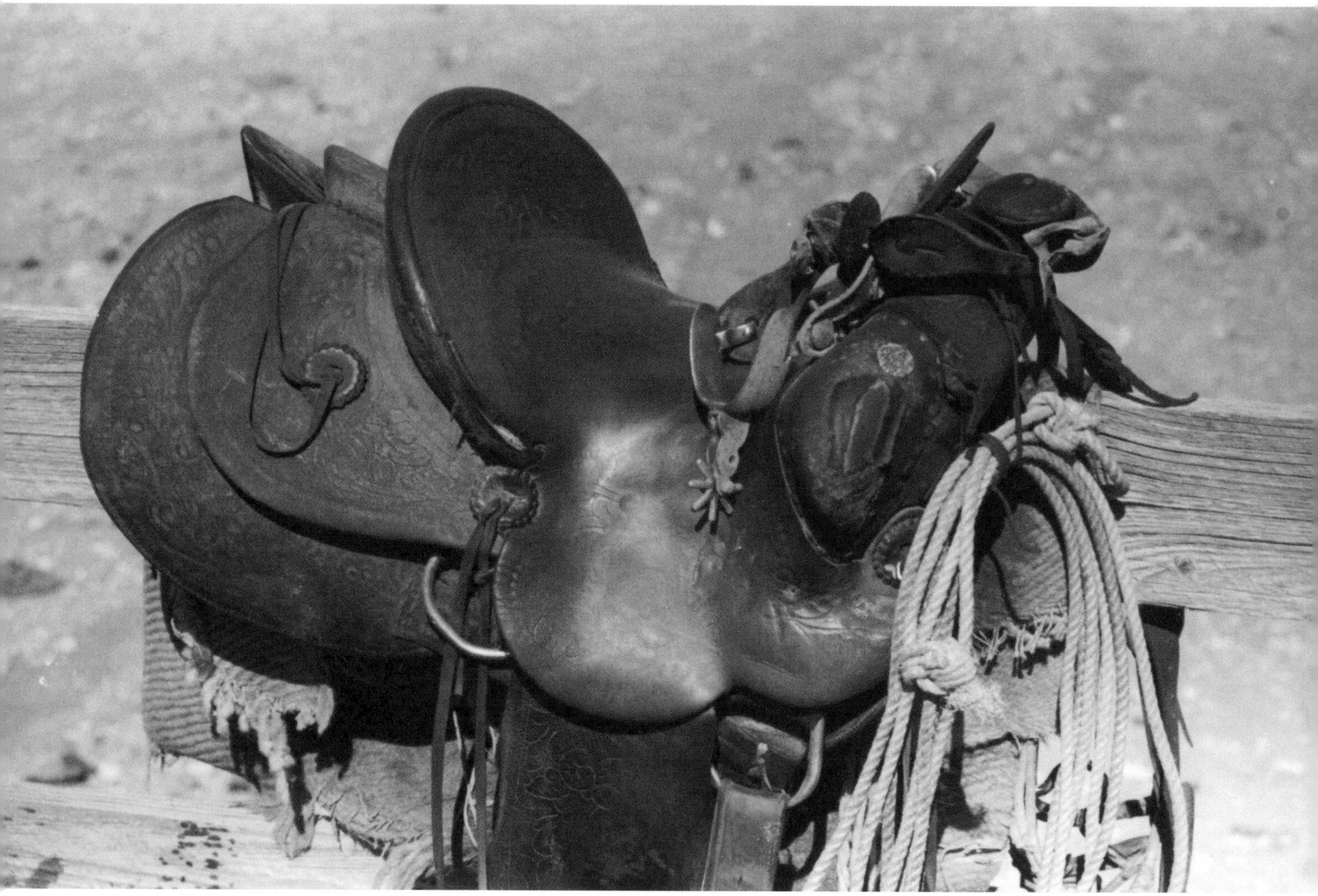

Detail of cowboy's saddle. Roundup near Marfa, Texas. Russell Lee, May 1939.
LC-USF33-012297-M4 DLC

INTRODUCTION
THE PHOTOGRAPHS OF THE FARM SECURITY ADMINISTRATION

I don't remember exactly when I first saw a Farm Security Administration (FSA) photograph, but I remember what it was very well. It was Dorothea Lange's 1936 "Migrant Mother" photo, and it transfixed me. Somehow Lange had captured not only the tragedy of poverty and suffering, but also the dignity and strength of the subject. That single photo powerfully represented the Depression, the Dust Bowl, and the desperate migration of people westward. Since then I have been fascinated by the work of the FSA photographers and their powers of observation and empathy with the lives of ordinary people. In 1979, I curated an exhibit in San Francisco of FSA photographs reflecting the folklife of California during the Depression. The sweeping scope of the photographers was stunning. They turned their cameras on Okie migrants, Portuguese religious celebrations, Mexican American farmworkers, social dances, rodeos, and more. In doing so, they created a portrait of the rich cultural diversity of the state in the late 1930s and early 1940s. Of particular interest to me was their representation of cowboy and ranch life, and that led me back to the Library of Congress to search out other FSA photos, not just from California but from the entire West. From the hundreds and hundreds of photos of cowboy and ranching culture, I have selected 125 that I think reflect the many aspects of ranching life at the time.

In 1937, the country was in the grip of the Great Depression. The plight of many Americans was made worse by the Dust Bowl, an ecological disaster that affected the Great Plains states from 1933 to 1936, eventually displacing more than 2.5 million people. In 1933, President Franklin D. Roosevelt initiated a series of programs that he called the "New Deal" to address the country's problems. One of these was the Resettlement Administration (RA), formed in 1935, an agency designed to aid in the resettlement of the displaced. It was succeeded in 1937 by the Farm Security Administration, part of the Department of Agriculture until it was finally moved to the Office of War Information. Both were New Deal programs intended to help farmers and ranchers suffering from the impacts of the Dust Bowl and the Great Depression.

Roy Emerson Stryker, a former economics professor at Columbia University, was placed in charge of a special photographic section of the RA and later of the FSA. He had grown up in Colorado, where as a young man he spent several years running a cattle ranch with his brother. During his time at Columbia, Stryker became intrigued by the use of photographs to illustrate social problems. His work in collecting and using photographs in the 1925 book *American Economic Life* and its subsequent editions made him a natural candidate for the FSA post.

Stryker enlisted the services of a number of the country's finest photographers, including Dorothea Lange, Russell Lee, John Collier, Arthur Rothstein, Marion Post Wolcott, John Vachon, Walker Evans, Ben Shahn, and Jack Delano. Stryker wanted photographers with an "insatiable curiosity, the kind that can get to the core of an assignment, the kind that can comprehend what a truck driver, or a farmer, or a driller, or a housewife thinks and feels and translate those thoughts and feelings into pictures that can be similarly comprehended by anyone." He told the photographers, "I want you to take pictures of everything you can find of what's happening to the people." Between 1935 and 1942 photographers working for Stryker took some 77,000 black-and-white photographs plus 644 color photographs. The Farm Security Administration/Office of War Information Photograph Collection is now housed at the Library of Congress.

The purpose of the project was twofold: to demonstrate the need for government assistance to the rural poor and to document the lives of everyday Americans on the farms, on the ranches, and in the small towns of rural America. Years later photographer Arthur Rothstein remembered, "There was a feeling that you were in on something new and exciting, a missionary sense of dedication to this project, of making the world a better place to live in." It was a time of great change and social upheaval in the country, encompassing the Depression, the Dust Bowl migrations, and the beginning of the Second World War and its social dislocations.

The collection includes some of the finest and most widely recognized documentary photographs ever taken. The noted photographer Edward Steichen, the editor of the famous 1955 *Family of Man* photo exhibition and book, wrote that these photos were "the most remarkable human documents ever rendered in pictures."

The best known of the FSA photographs were of Dust Bowl immigrants in Oklahoma and California, Depression-era soup lines, and farm life in states such as Vermont and Kentucky, but the FSA photographers also visited the ranching country of the rural West. A number of photographers had previously documented the cowboys and cattle culture of their time, including Montana photographer L.A. Huffman during the late 1800s, Erwin Smith, who was active in Texas and the Southwest in the early 1900s, and Charles J. Belden in 1920s Wyoming. But the period recorded by the FSA photographers, 1935 through 1943, was an important time of transition. Quite a few of the older generation of cowboys who had "gone up the trail" with herds of cattle in the late 1800s and early 1900s were still alive, but times were rapidly changing. The advent of barbed wire in the late 1800s signaled the end of the open range and long trail drives. Cowboys were now working on ranches circumscribed by fences. Modern technologies, such as automobiles, trucks, and tractors, were changing the way work was done. Cowboys still tended cattle on horseback, but rather than riding for hours to their herds, they could put their mounts in a horse trailer and drive to where they were needed. Livestock could now be transported in trucks. Hay could be cut and hauled by tractor. In many rural areas people now had electric lights and appliances and got their news and entertainment from the radio. They talked with each other by telephone.

It is this important period of change that was captured by the FSA photographers. The photographers selected by Stryker had various backgrounds and strengths, but all had a keen awareness and used their cameras to document many different aspects of life in the communities they visited. They documented cowboys at work, but they also looked at the everyday lives of ranching women and children. They photographed livestock auctions, dances, people cooking and eating and worshipping. A photograph of the interior of a bunkhouse, for example, or a café, or a schoolhouse, could tell much about the lives of those who lived there. And, importantly, they captured evidence of change, as in Russell Lee's photograph of a cook making bread at a traditional chuck wagon with a couple of automobiles parked nearby. Or Arthur Rothstein's photos of a Montana cowboy documenting his own surroundings with a home movie camera, or of a ranch wife talking on a party line on an early crank telephone. The result is an amazingly rich and personal record of ranch life of the period. In selecting the photographs for this book, I have chosen those that present the broadest possible view of cowboy and ranch life. These photos carry us back in time to a specific moment and place, but they also bring us intimately and immediately face-to-face with the humanity of the subjects. I hope you find them as compelling as I do.

—CHARLIE SEEMANN

RUSSELL LEE 1903–1986

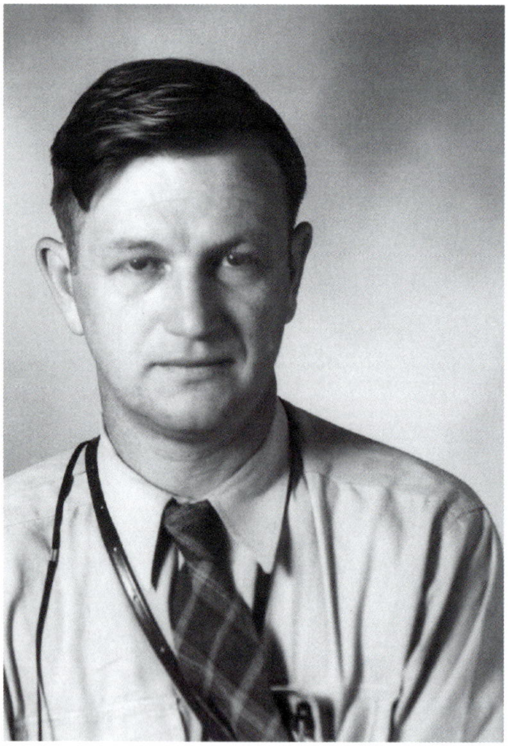

Portrait of Russell Lee, 1942.
LC-DIG-fsa-8d26773

Russell Lee was born in 1903 in Ottawa, Illinois. During the fall of 1917, Lee enrolled in Culver Military Academy in Indiana, where he spent the next four years, graduating in 1921. Lee then attended Lehigh University in Pennsylvania from 1921 to 1925 and earned a degree in chemical engineering. By 1929, he had become disenchanted with his career and moved to San Francisco where he pursued painting. A little over a year later, upon the suggestion of a friend, he relocated to an artist's community in Woodstock, New York. He first used the camera to make studies for his paintings, but soon discovered he liked telling photographic stories more than he did painting. When an FSA staff photographer quit, Roy Stryker hired Lee to freelance a story on a homestead housing project in New Jersey. Stryker liked the photographs so much that he hired Lee full time. Lee stayed with the FSA from 1936 to its demise in 1943. He was among the most prolific of the FSA photographers, spending much of this time photographing in and around Pie Town, New Mexico, and in the cattle country of central and west Texas. Lee was also one of the photographers who documented the evacuation of Japanese Americans from the West Coast to detention camps.

After the FSA was defunded in 1943, Lee served in the Air Transport Command (ATC), during which he took photographs of all the airfield approaches used by the ATC to supply the Armed Forces in World War II. He worked for the United States Department of the Interior (DOI) in 1946 and 1947, helping the agency compile a medical survey in the communities involved in mining bituminous coal. He took more than 4,000 photographs of miners and their working conditions in coal mines. In 1946, Lee completed a series of photos focused on a Pentecostal Church of God in a Kentucky coal camp. While completing the DOI work, Lee also continued to work under Stryker, producing public relations photographs for Standard Oil of New Jersey. Some 80,000 of those photographs have been donated by Exxon Corporation to the University of Louisville in Kentucky. In 1947 Lee moved to Austin, Texas, and continued photography. In 1965 he became the first instructor of photography at the University of Texas.

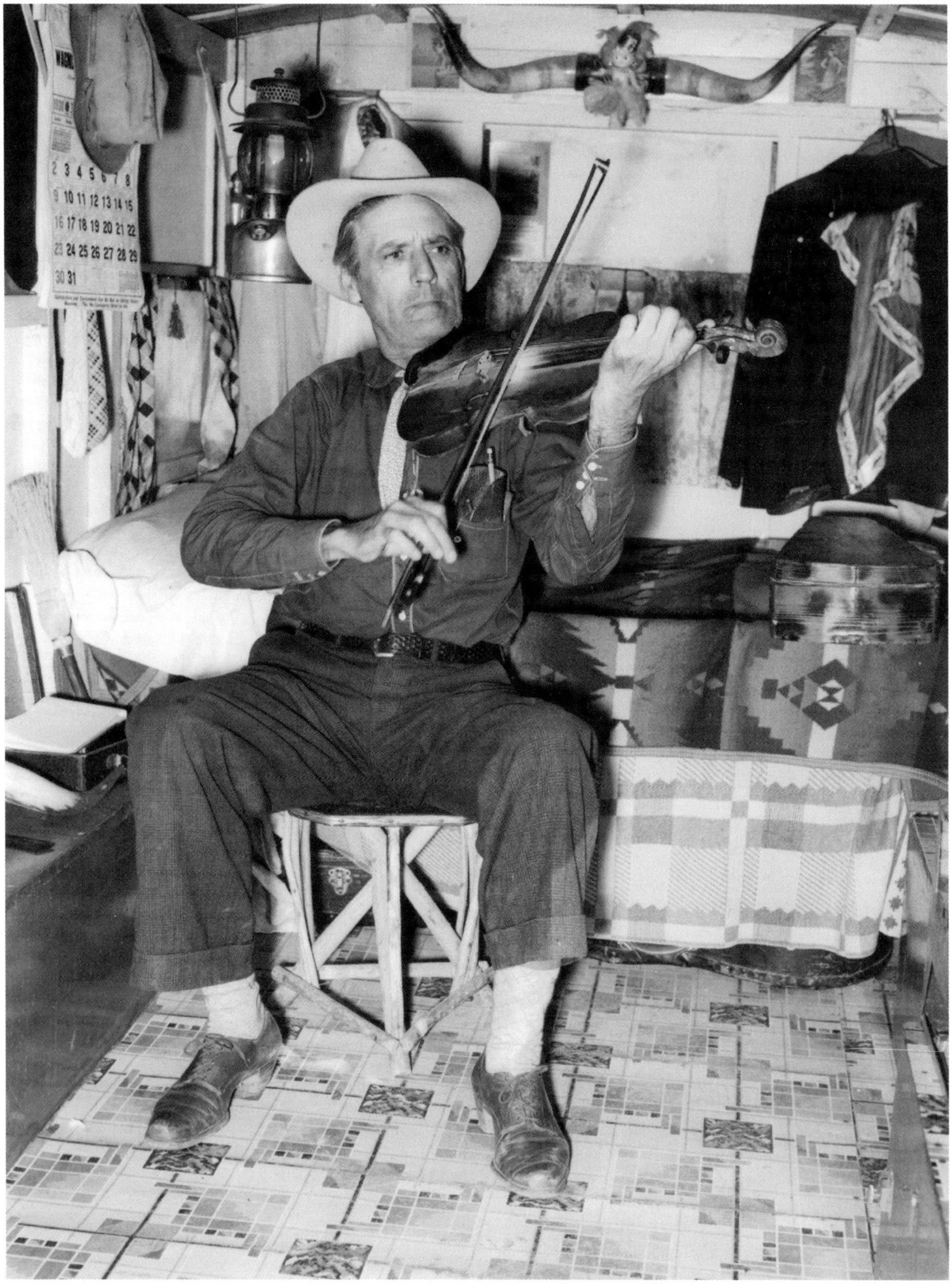

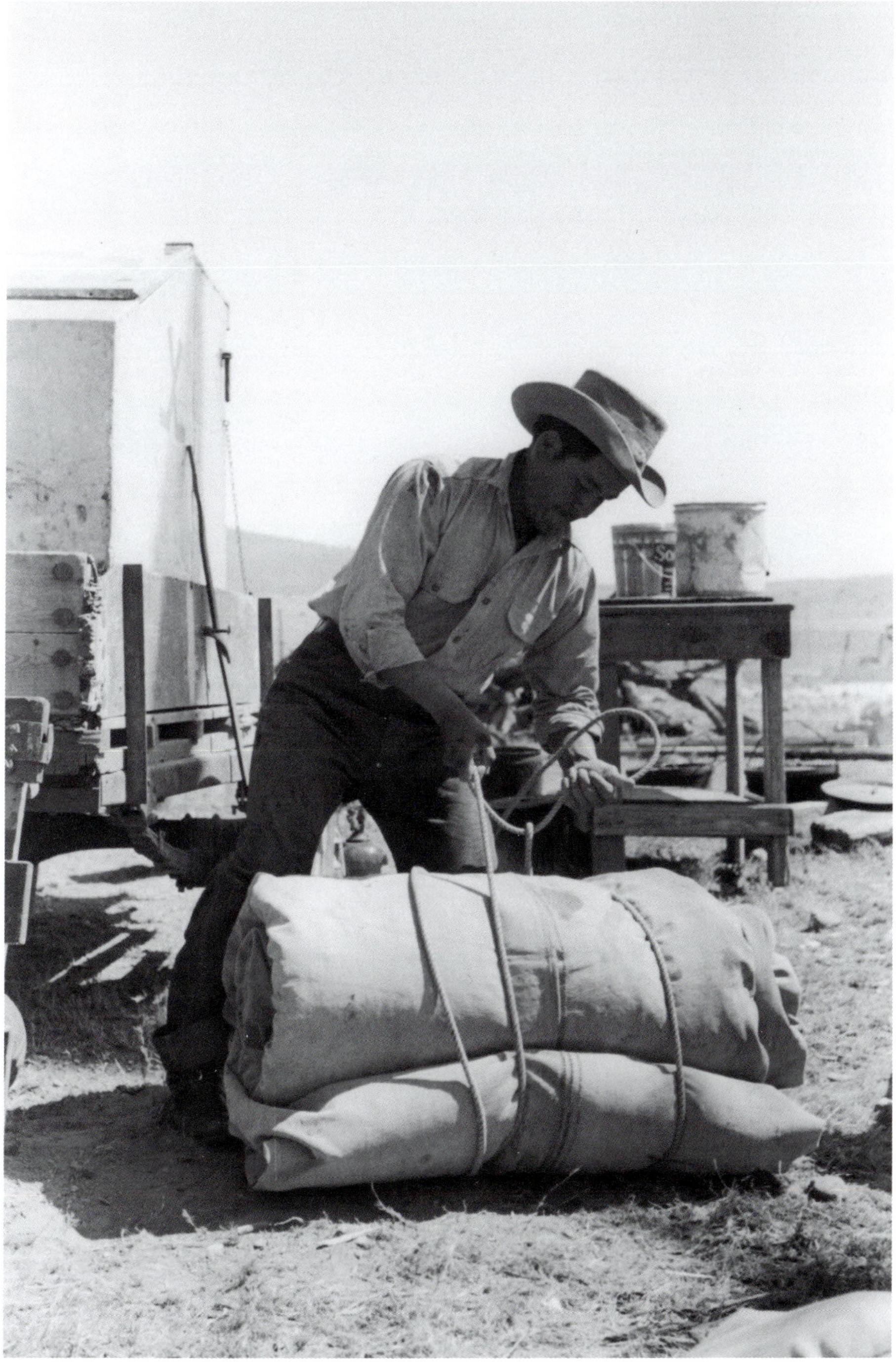

OPPOSITE: Mr. Bias, former cowboy, travels around the country in a trailer. Has private income. Weslaco, Texas. Russell Lee, February 1939.
LC-USF34-032110-D DLC

LEFT: Cattle ranch near Marfa, Texas. Cowboy tying up bedroll. Russell Lee, May 1939.
LC-USF33-012210-M4 DLC

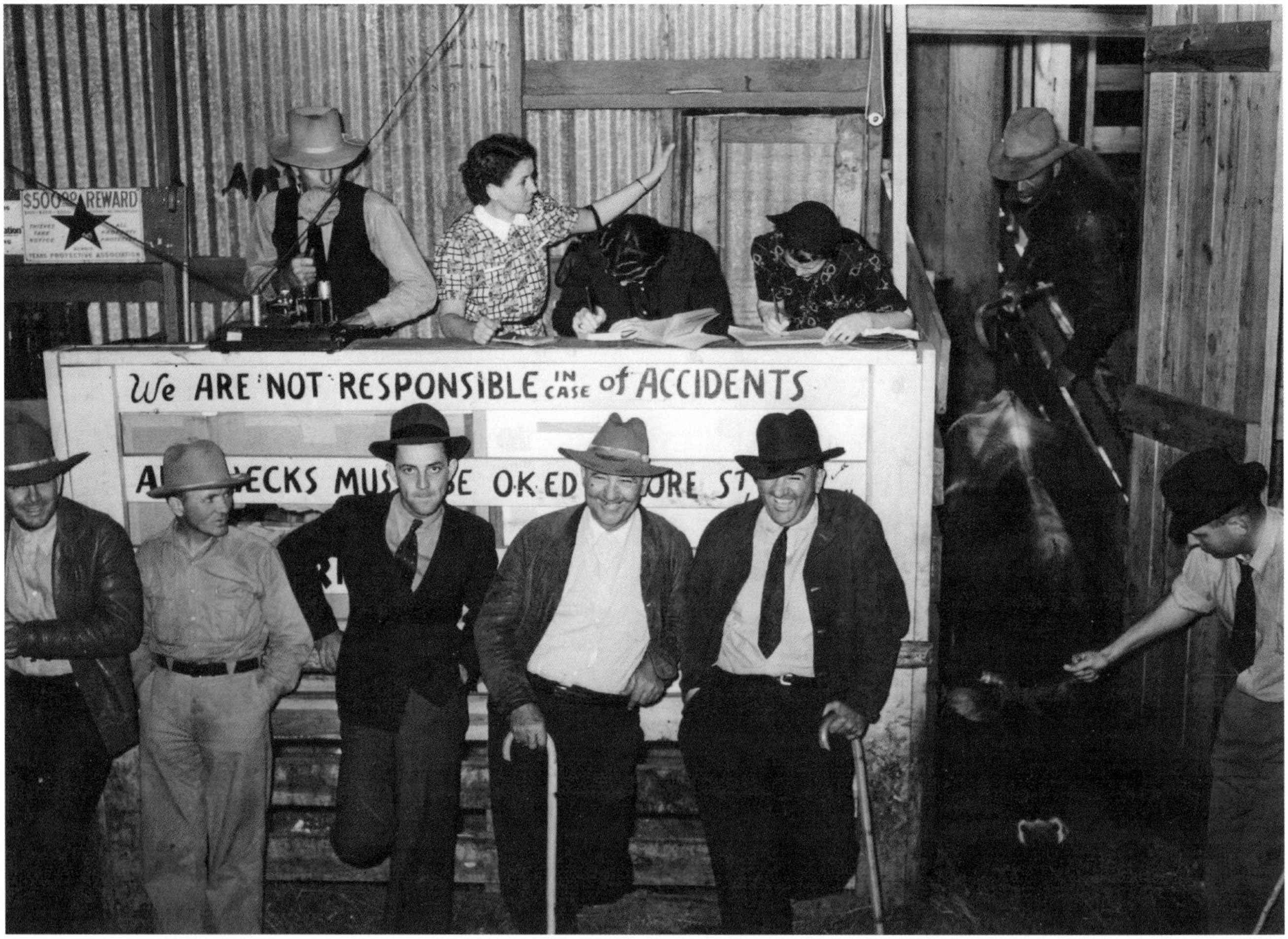
REWARD
We ARE NOT RESPONSIBLE IN CASE of ACCIDENTS

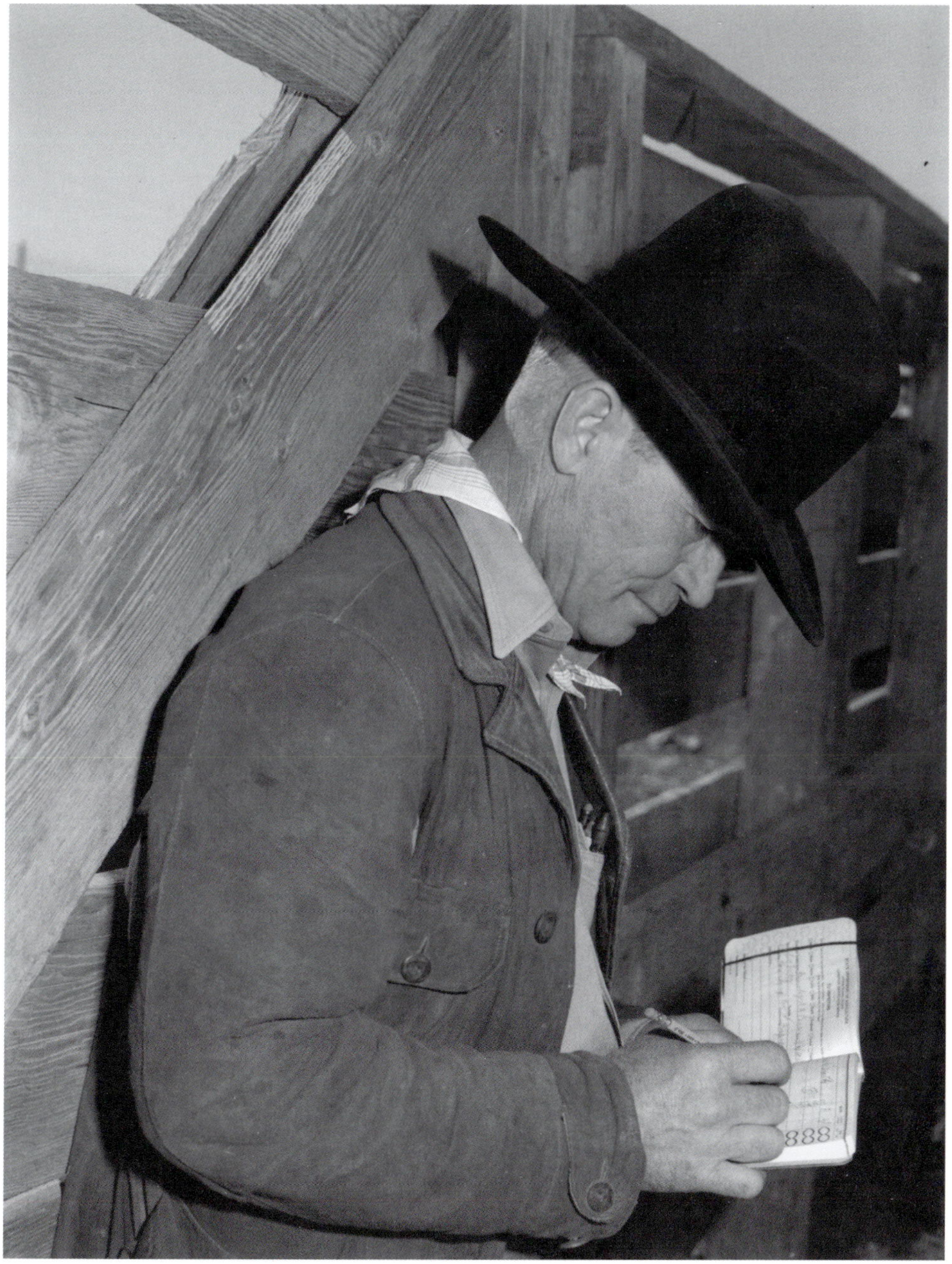

OPPOSITE: Auction scene in San Augustine, Texas. Cow is being brought in from the pen. The auctioneer is using a public-address system. Russell Lee, April 1939. *LC-USF34-032924-D DLC 5*

LEFT: Cattleman notes down the number of head of cattle shipped. Imperial County, California. Russell Lee, March 1942. *LC-USF34-072025-D*

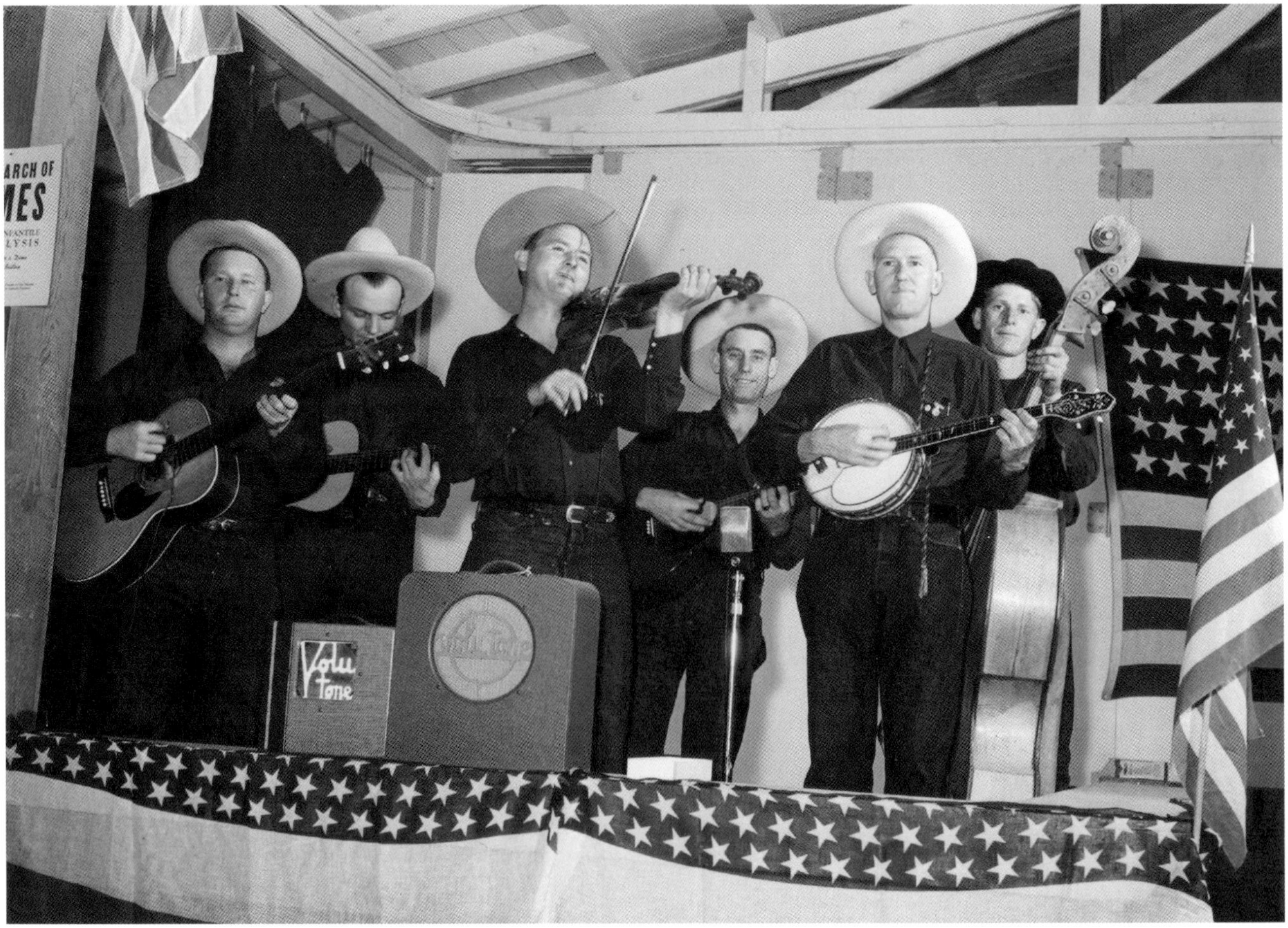
Volu tone

MEXICAN COWBOYS

Although most popular depictions of American cowboys portray them as white, in fact working cowboys during the cattle drive days of the late 1800s were a diverse lot, including Anglos, African Americans, Mexicans, and Native Americans. It has been estimated that in Texas, Arizona, and New Mexico, lands that were previously part of Northern Mexico, as many as one in three cowboys was Mexican or Mexican American. Many of these *vaqueros* were native to these regions, from towns and villages on both sides of the border. They were an important source of ranch labor in the 1930s and 1940s, as they are in these regions to this day.

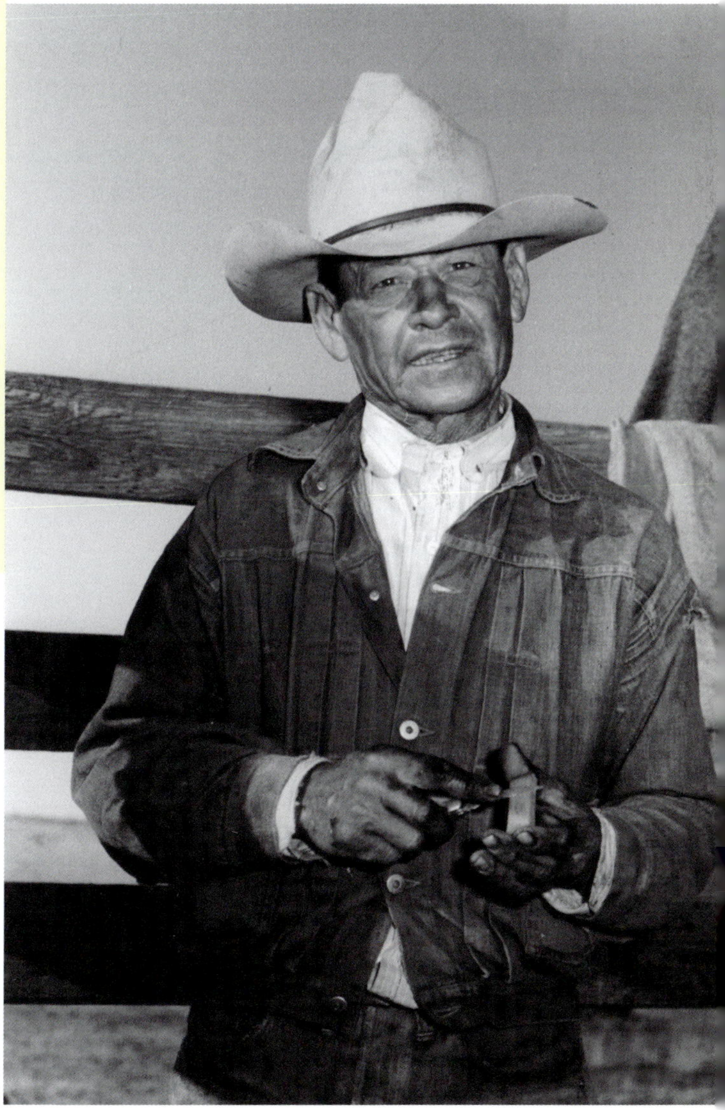

OPPOSITE: Farm Security Administration farmworkers' camp in Tulare County, California. Hired orchestra that played at the president's birthday ball. Russell Lee, February 1942.
LC-DIG-fsa-8c23804 DLC; LC-USF34-071665-D DLC

RIGHT: Roundup near Marfa, Texas. Mexican cowboy sharpening his knife. Russell Lee, May 1939.
LC-USF33-012294-M2

WATER WITCHING

BELOW: Water witch. "Green forked stick is held firmly thus." Pie Town, New Mexico. Russell Lee, June 1940. *LC-USF33-012750-M2 DLC*

OPPOSITE: Mexican boy playing guitar in room of corral. Robstown, Texas. Russell Lee, February 1939. *LC-DIG-fsa-8b20985 DLC: LC-USF34-032300-D DLC*

During the drought years of the Dust Bowl, many ranchers and farmers turned to "rain makers" and "water witches" in their desperate search for rain and ground water. Water witching, or dowsing, is a type of divination used in attempts to locate ground water, usually using a Y- or L-shaped twig or rod, called a "witching rod" or a "divining rod." The practice, which science has long debunked, has a long history and may have originated in Germany as early as the fifteenth century. Even today there are folks who seek the help of water witches in locating sites for wells. A great description of water witching during the Dust Bowl is found in Lois Phillips Hudson's short story "The Water Witch," in her book *Reapers of the Dust: A Prairie Chronicle.* She wrote that "desperate hunts for water went on over millions of square miles of the breadbasket that was becoming the Dust Bowl in the fourteen summers of desolation."

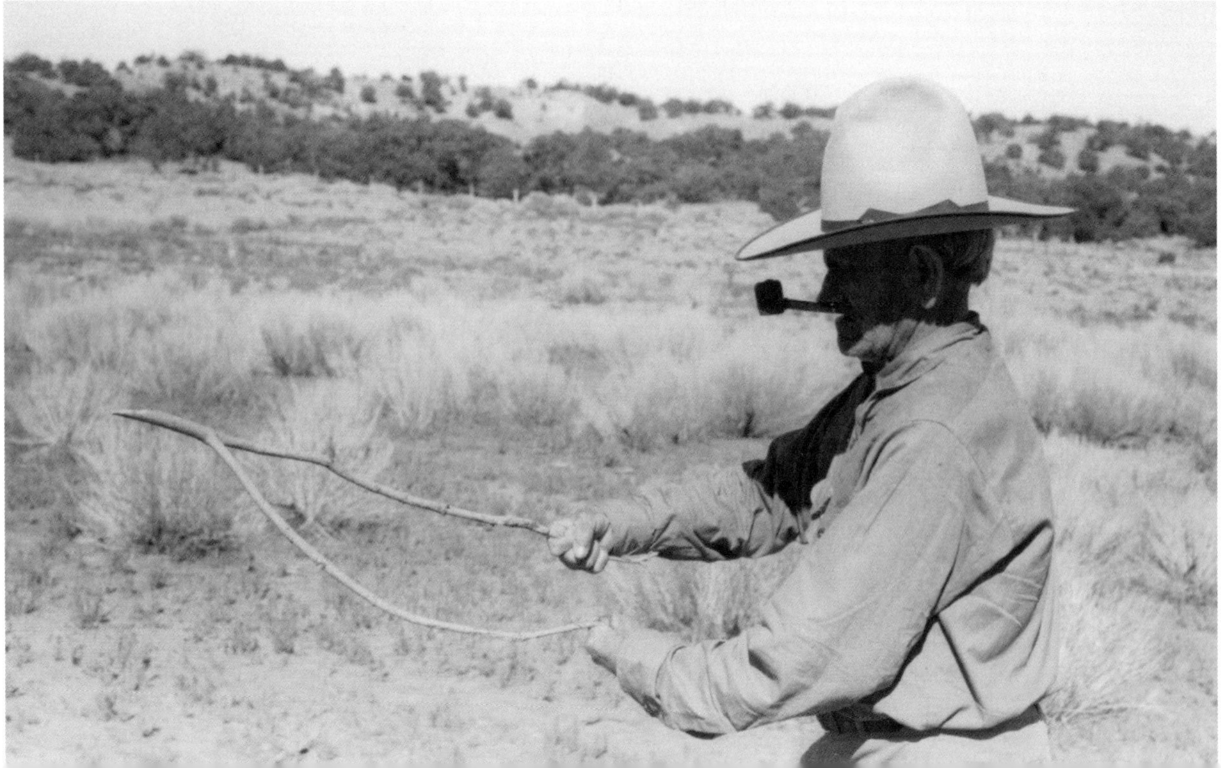

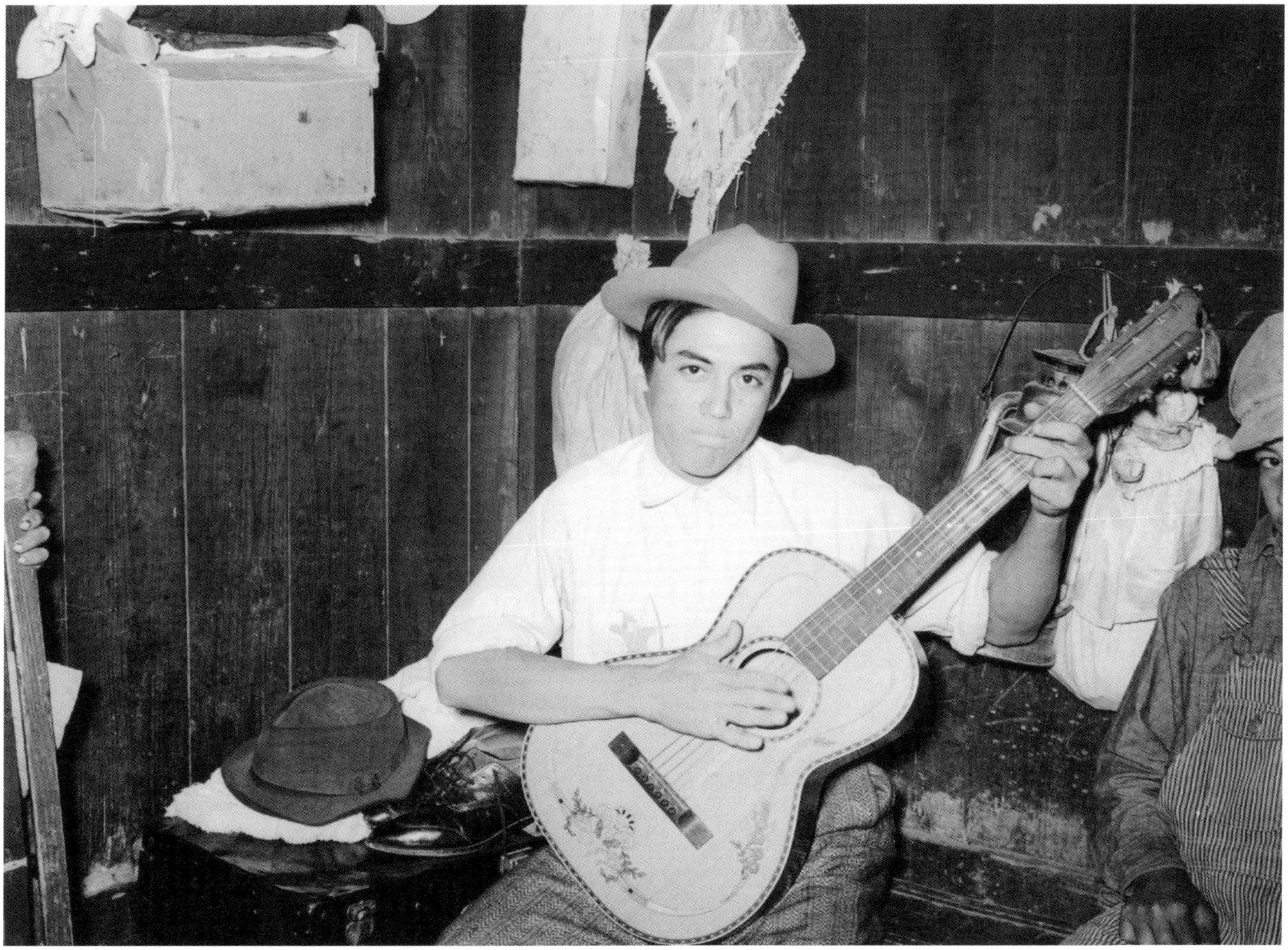

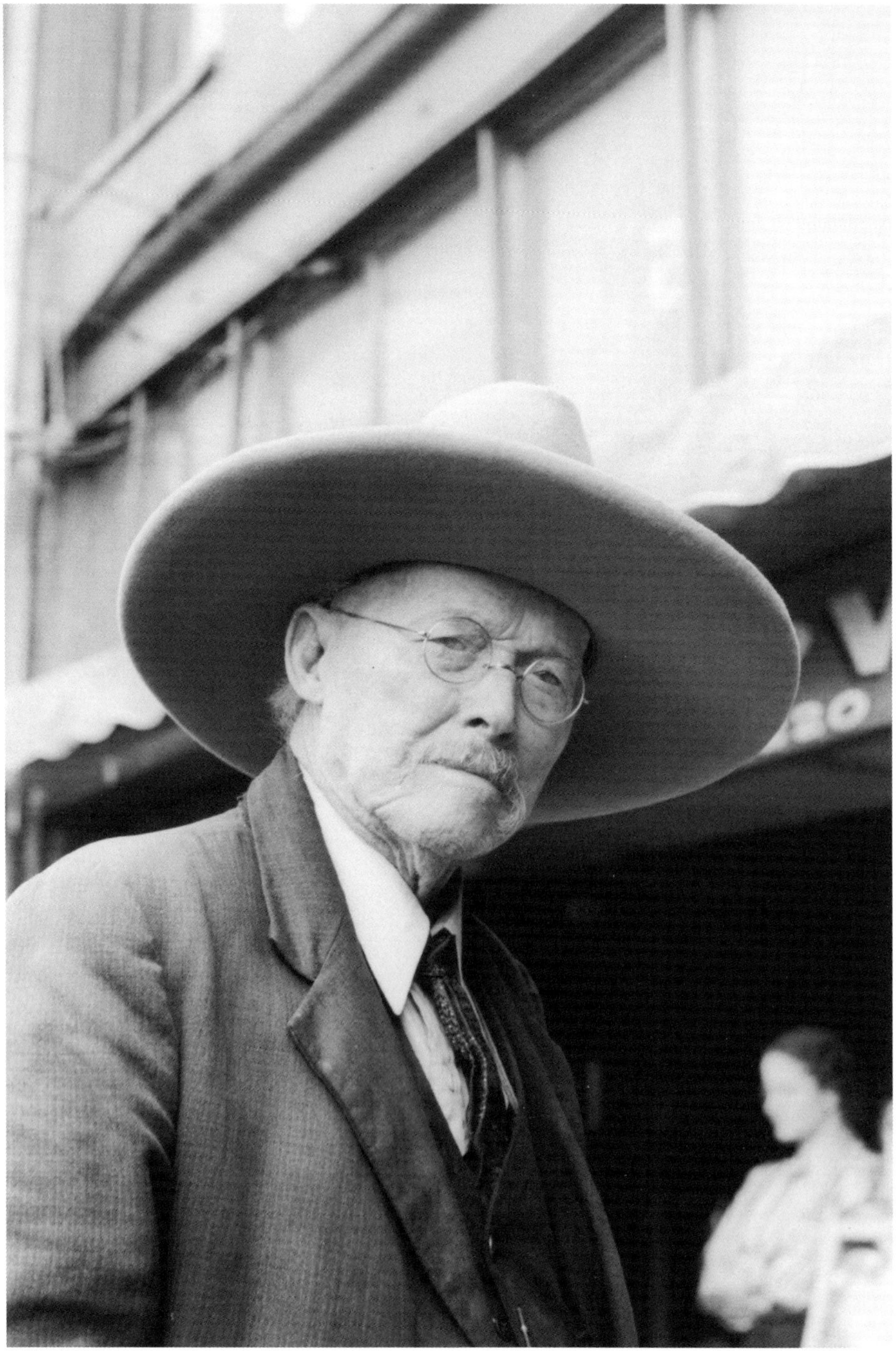

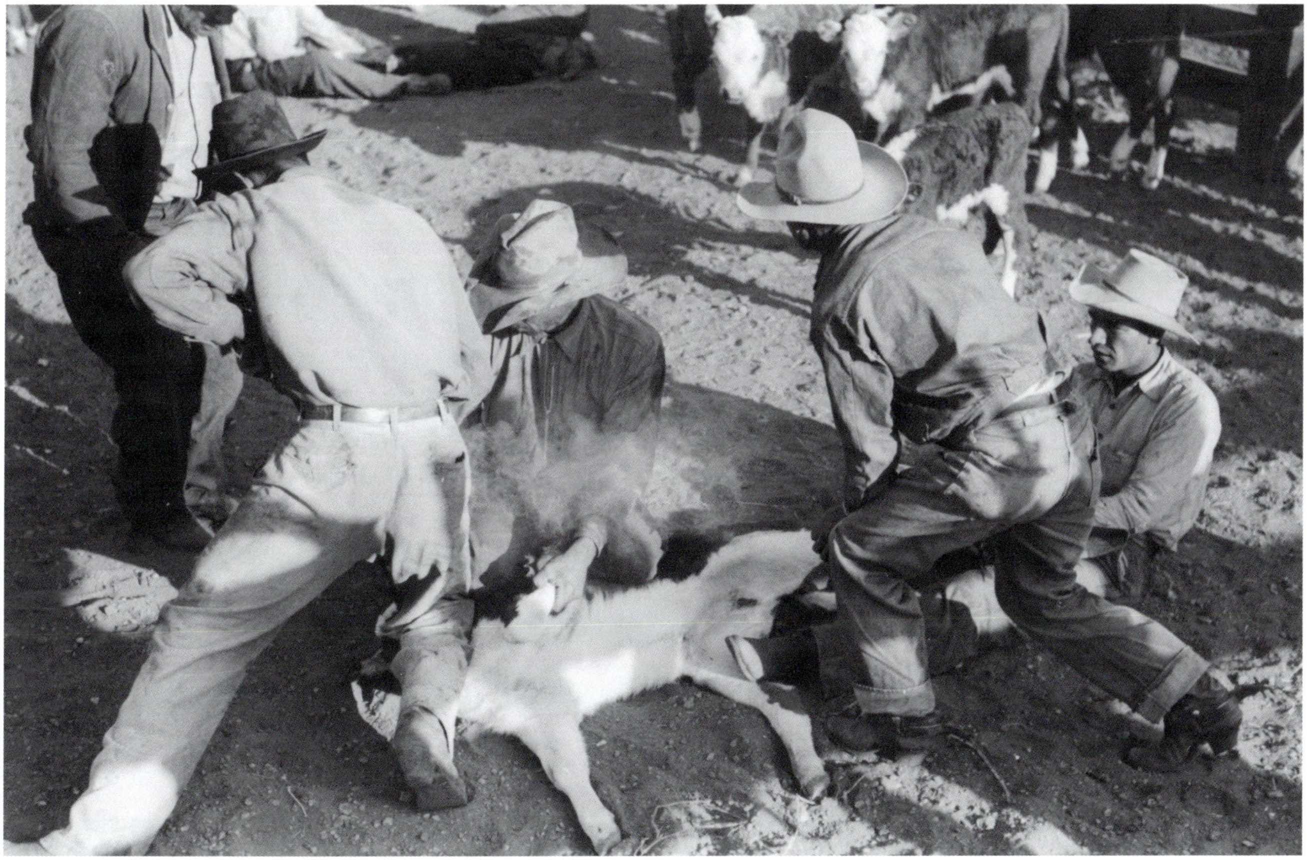

OPPOSITE: Texan. Waco, Texas. Russell Lee, November 1939.
LC-USF33-012487-M3 DLC

ABOVE: Working on a calf during branding operations at roundup near Marfa, Texas. Russell Lee, May 1939.
LC-USF33-012290-M5 DLC 13.

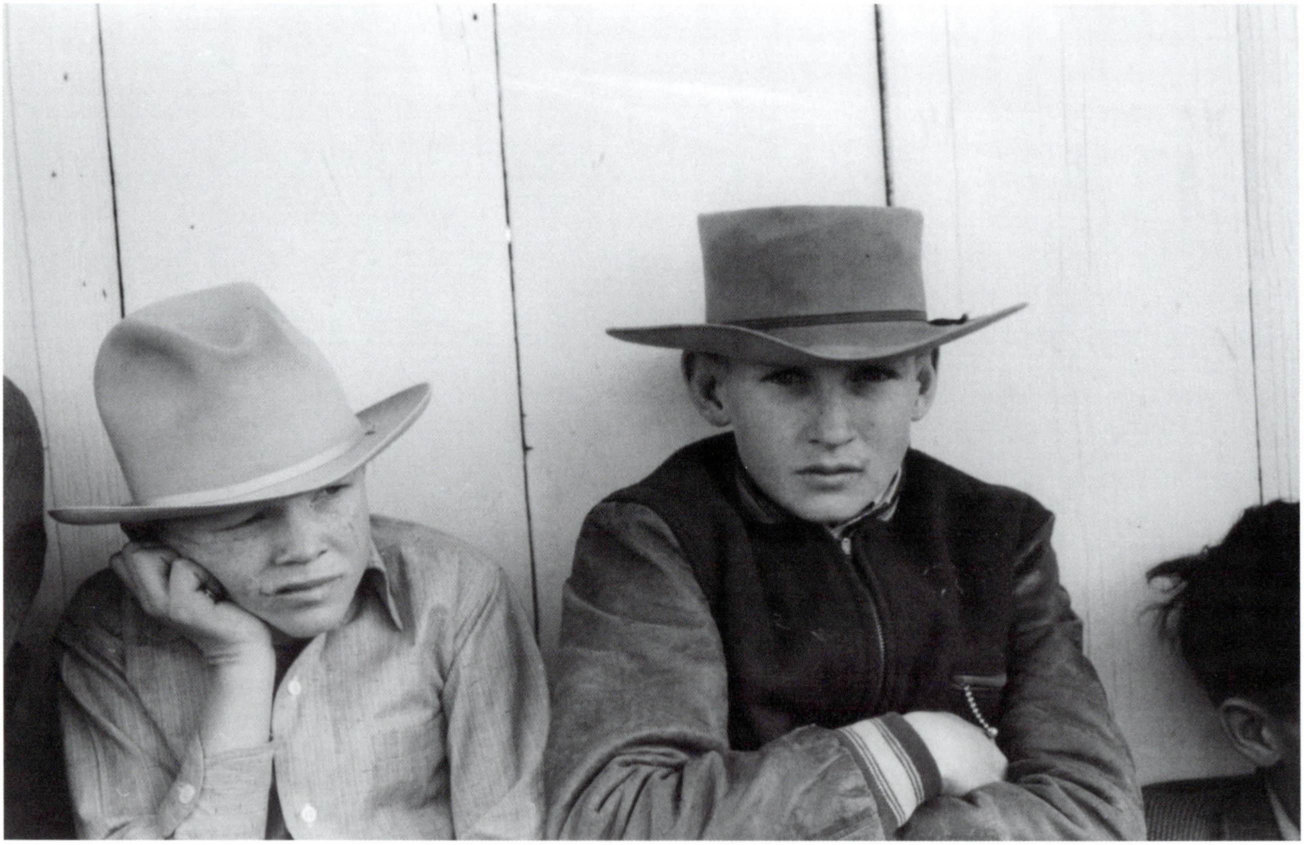

ABOVE: West Texas boys at the San Angelo Fat Stock Show. San Angelo, Texas. Russell Lee, March 1940.
LC-USF33-012595-M4

OPPOSITE: Member of an orchestra that plays outside a grocery store on Saturday afternoon tuning his violin. Phoenix, Arizona. Russell Lee, May 1940.
LC-USF33-012649-M3 DLC

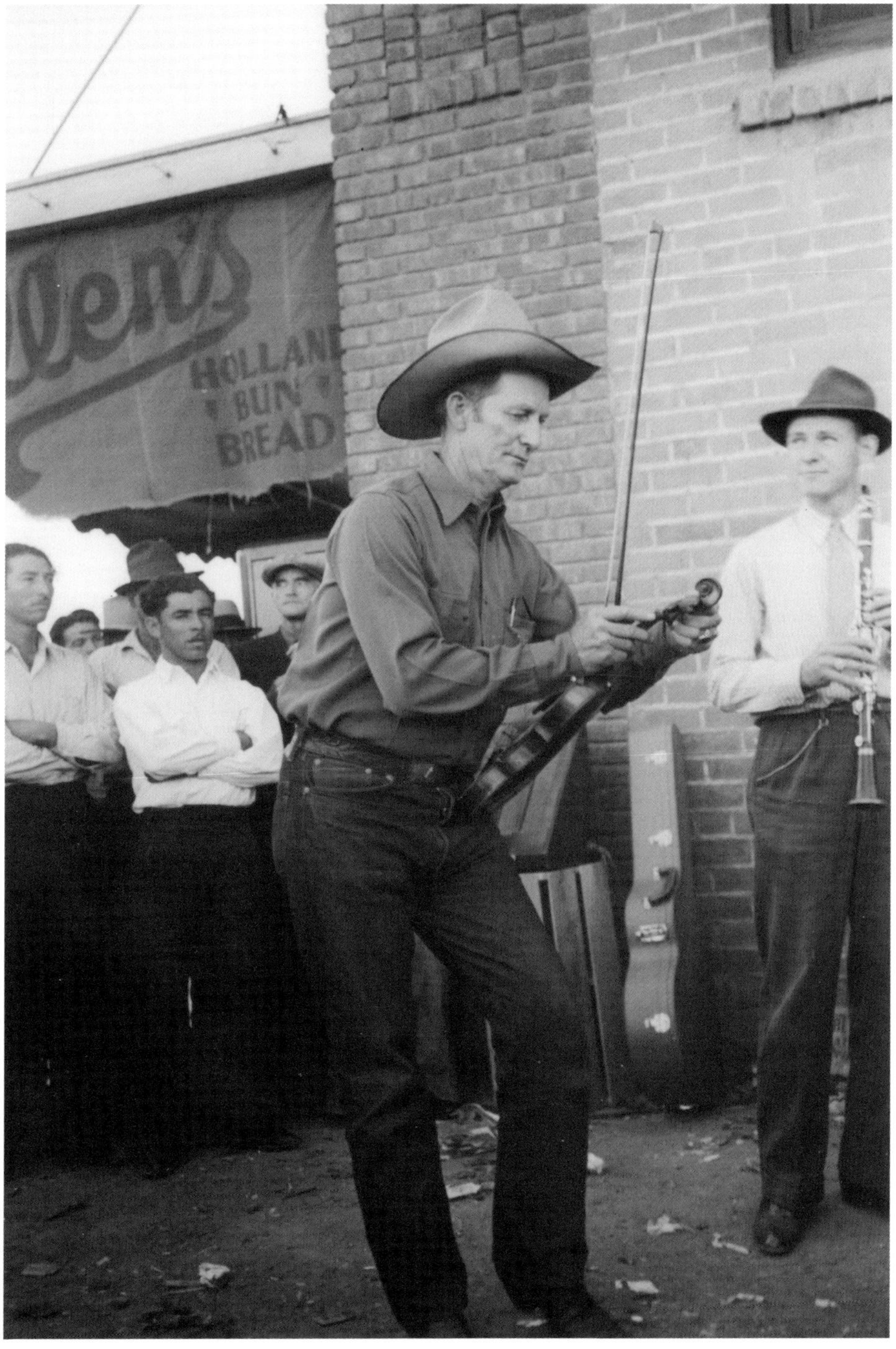
HOLLAND
BUN
BREAD

THE WASHING MACHINE

Few inventions have done as much to lighten the workload of women through the ages as the washing machine. Probably the earliest washing "machine" was the scrub board, invented in 1797. American James King patented the first washing machine to use a drum in 1851, but it was still hand-powered. In 1858, Hamilton Smith patented the rotary washing machine. The first wringer was added to a washing machine in 1861. In 1874, William Blackstone of Indiana built a machine that removed and washed away dirt from clothes. These first machines were often hand-operated. Some early machines were operated by a low-speed single-cylinder gasoline engine. The first electric washing machine was invented around the late 1800s to early 1900s, but electricity was not commonly available until at least 1930. In 1928 the Hurley Machine Company of Chicago, Illinois, introduced an electric-powered washing machine called "The Thor." Electric washing machine sales in the United States reached 913,000 units in 1928. However, high unemployment rates in the Depression years hit sales hard and by 1932 the number of units shipped was down to about 600,000. By 1940, sixty percent of the 25,000,000 wired homes in the United States had an electric washing machine. With the coming of electricity to the ranch, the electric washing machine was not far behind. Washing machines saved women twenty eight-hour days a year scrubbing the family laundry.

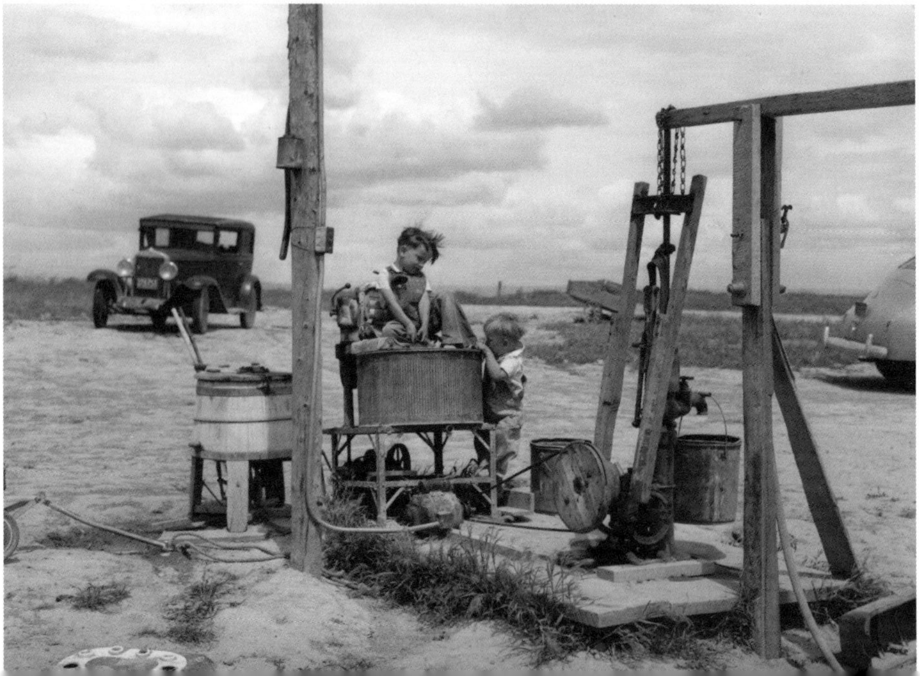

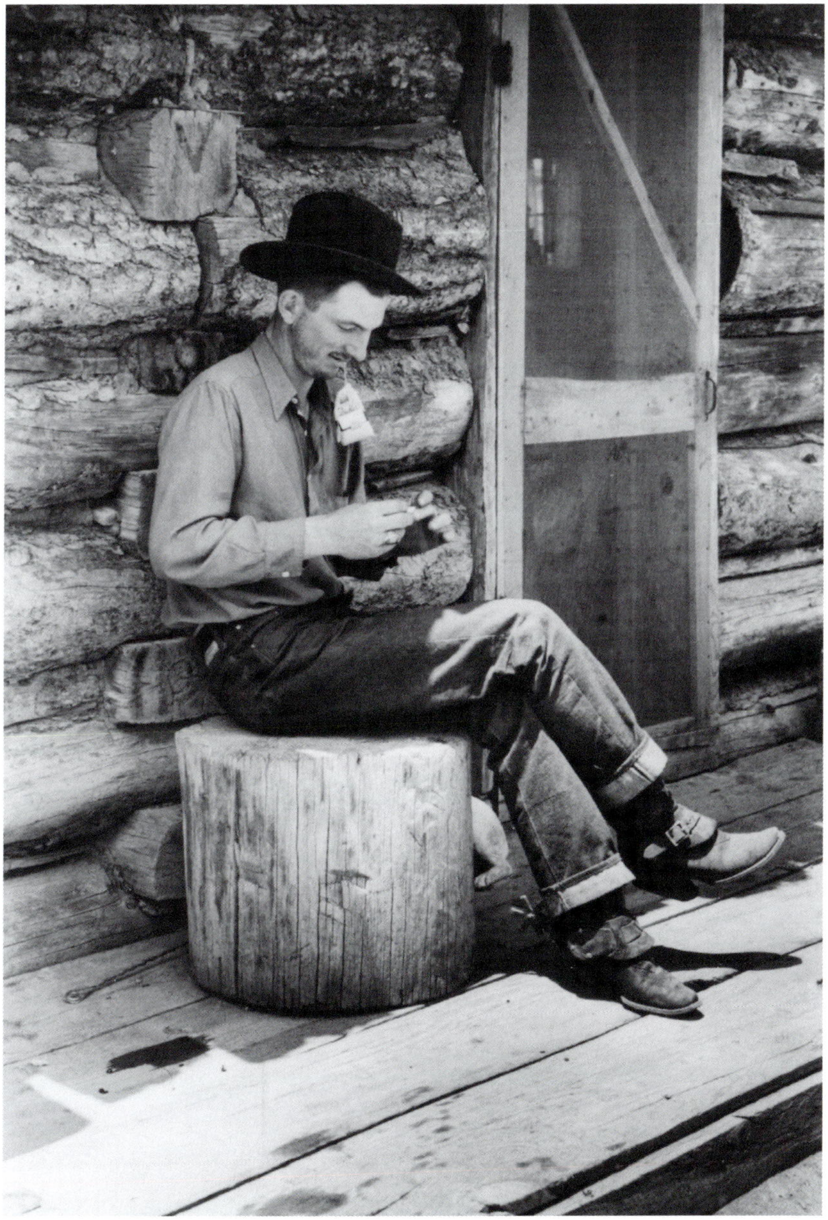

OPPOSITE: Children of Ray Halstead playing around the well. An electric motor is arranged so that it pumps water and supplies power for the washing machine. Dead Ox Flat, Malheur County, Oregon. Russell Lee, May 1941. *LC-USF34-038946-D DLC*

LEFT: A farmer rolling a cigarette. The construction of a log house is shown. This farmer rents the farm and house from the man who homesteaded the place but who is now unable to farm. Pie Town, New Mexico. Russell Lee, June 1940. *LC-USF331-012733-M1 DLC*

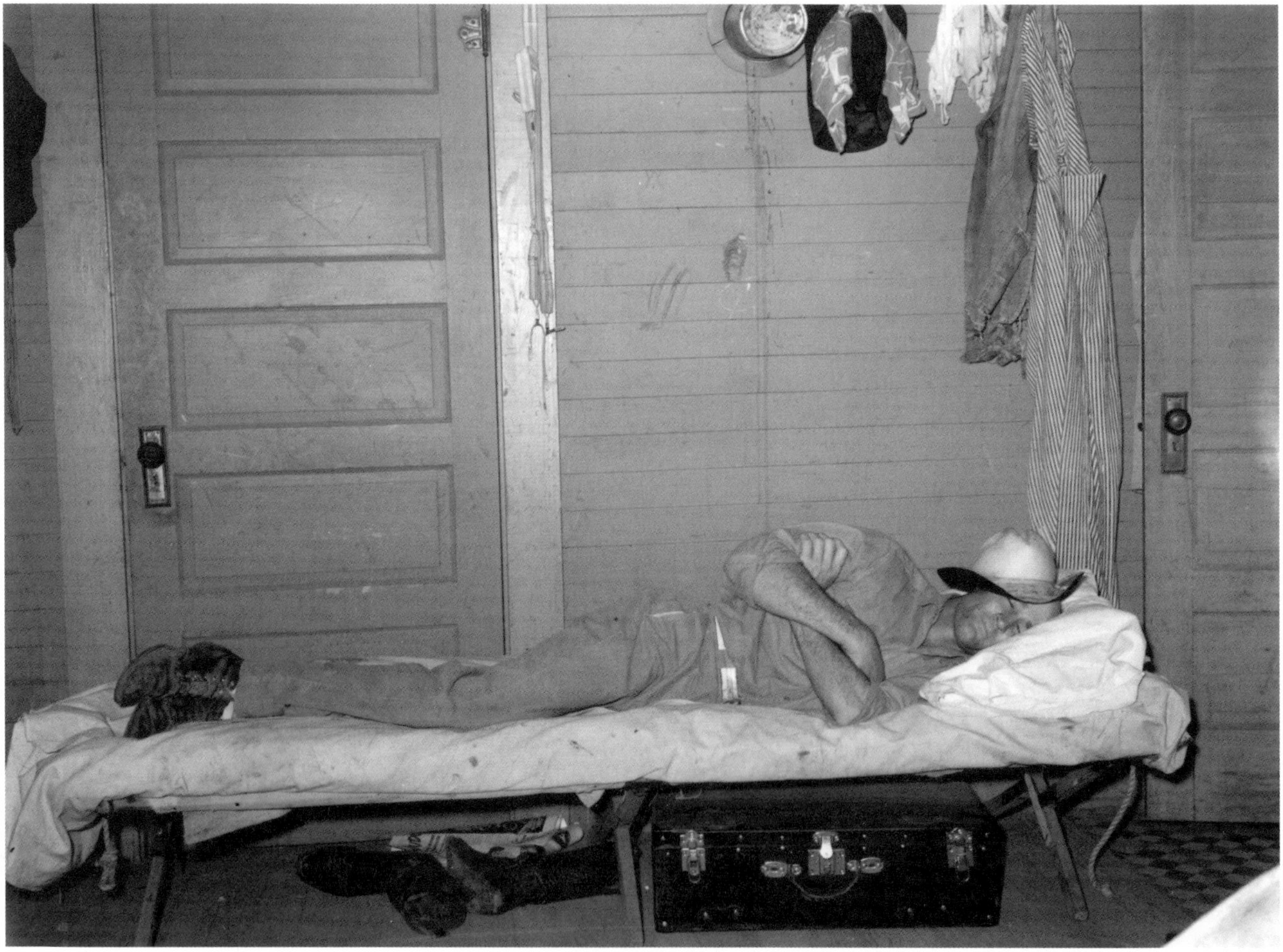

ABOVE: Cook of the SMS Ranch asleep after dinner. Near Spur, Texas. Russell Lee, May 1939.
LC-USF34-033339-D DLC.

OPPOSITE: Cowboys in town. San Juan Bautista, California. Russell Lee, May 1942.
LC-USF34-072632-E DLC

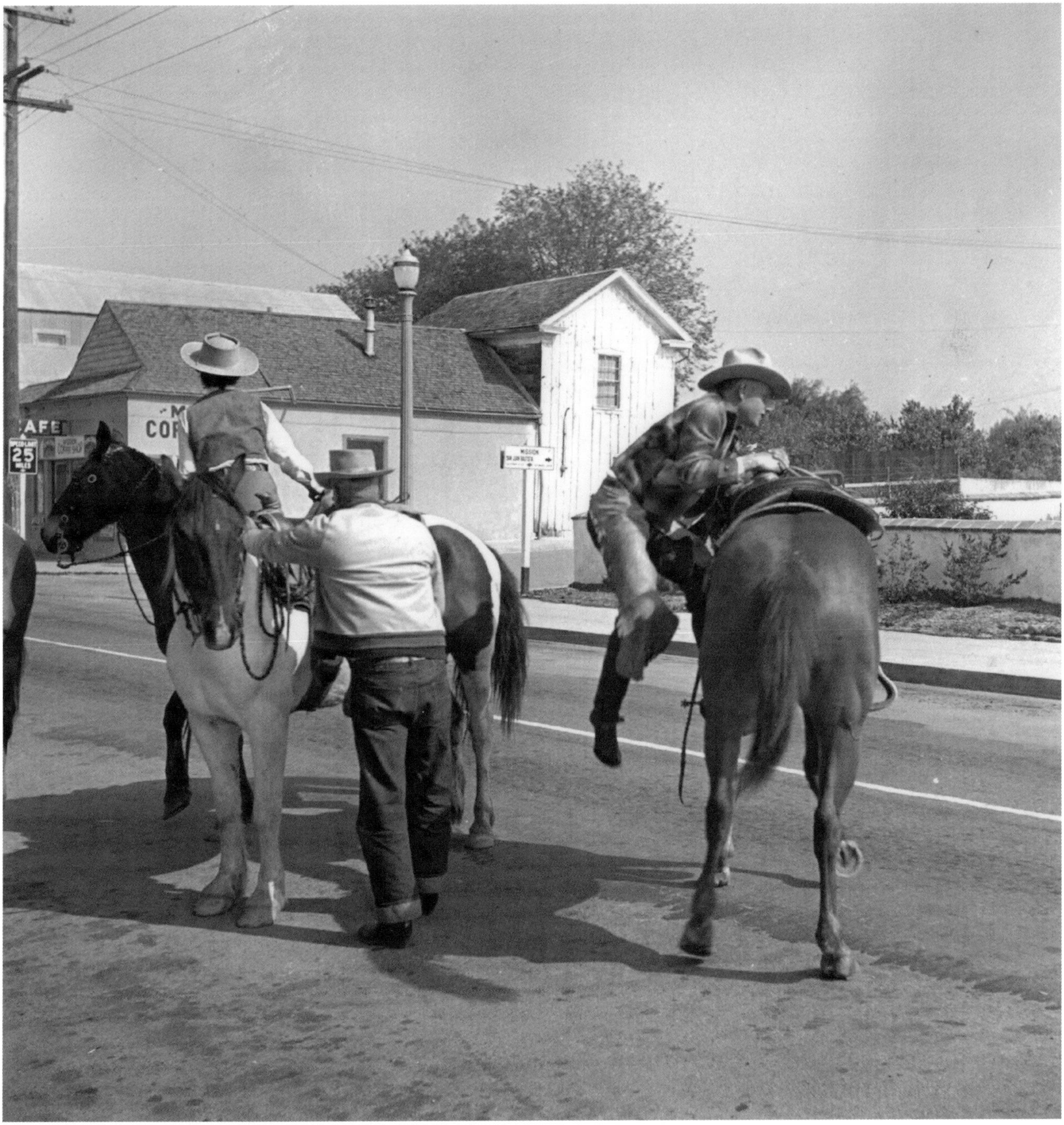
25
MISSION

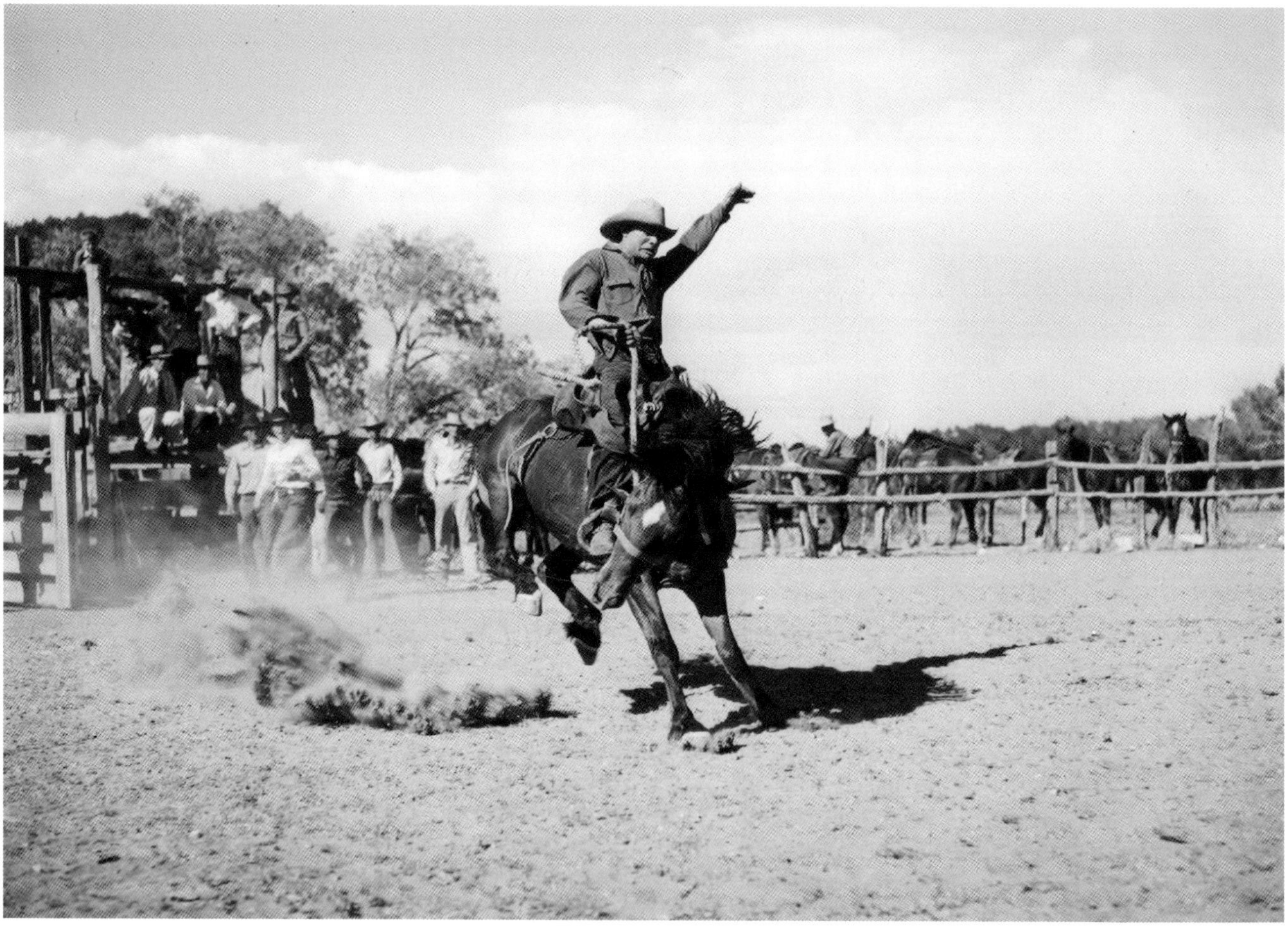

Bronc busting at the rodeo. Quemado, New Mexico. Russell Lee, June 1940.
LC-USF34-036948-D DLC

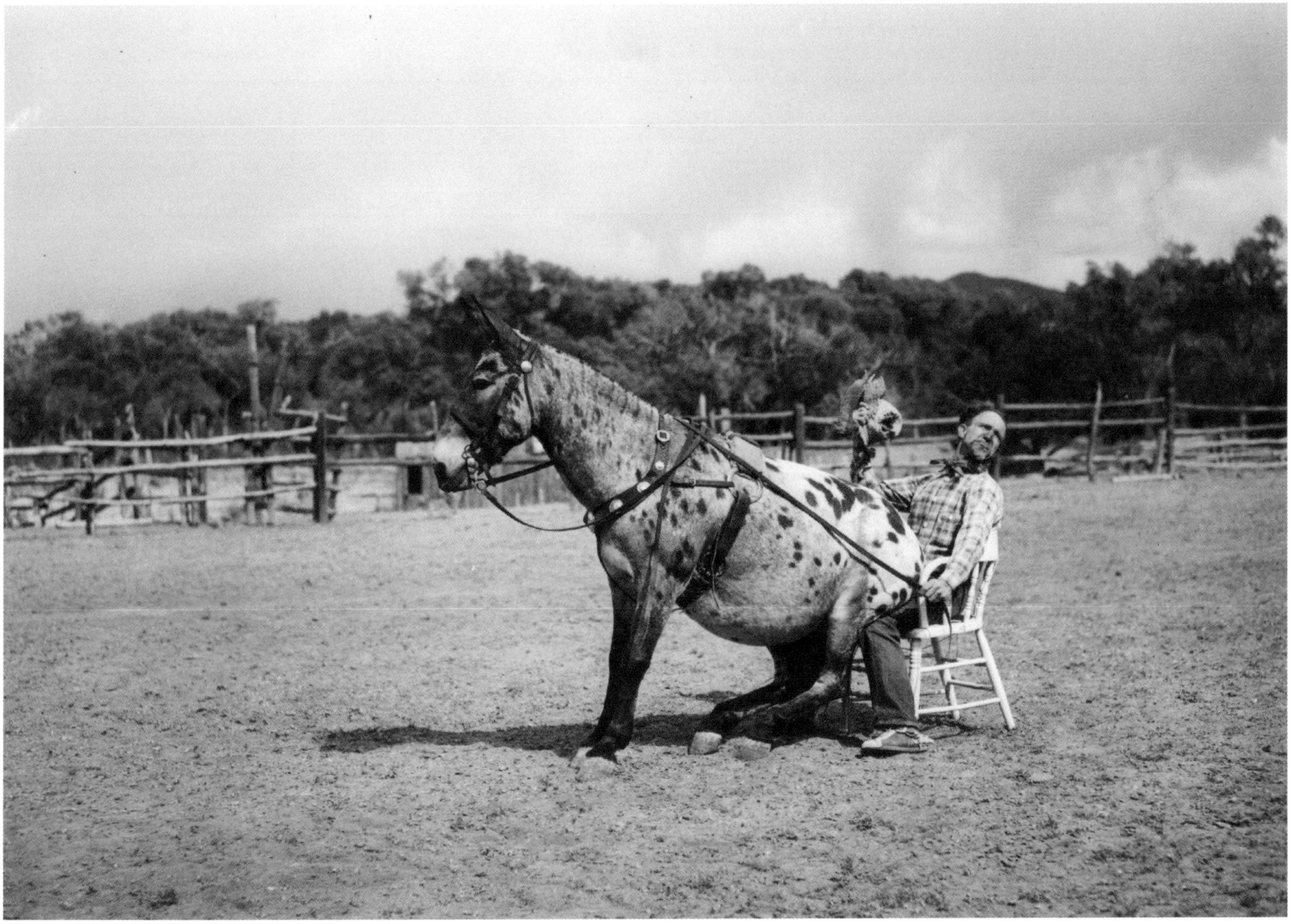

Clown rider with his trick mule at a rodeo. Quemado, New Mexico. Russell Lee, June 1940.
LC-USF34-036949-D DLC

THE COWBOY BOOT

The cowboy boot had its origins in the military boots of European horse soldiers, such as the Wellington boot, named for the Duke of Wellington. Americans wore this kind of riding boot, but after the Civil War, the style began to change. As cowboys began working the long trail drives from Texas to the railheads in Kansas, they asked boot makers for a more pointed toe so the boot could slip more easily into a stirrup and a higher heel so their foot couldn't slip through the stirrup. By the mid-1800s, boot makers in Kansas and Texas were creating tough, well-made working boots, and by the 1880s, the tall stovepipe top had been added.

Early boot makers included the Hyer Boot Company, founded about 1880 by brothers Charles and Edward Hyer in Olathe, Kansas, and Justin Boots, started in 1879 by H.J. Justin in Spanish Fort, Texas. They were followed by others such as Gus Blucher, who opened a shop in Cheyenne, Wyoming, in 1915 and Tony Lama in El Paso, Texas, in the 1930s.

It has been estimated that there are more than 370 steps in crafting a hand-made boot.

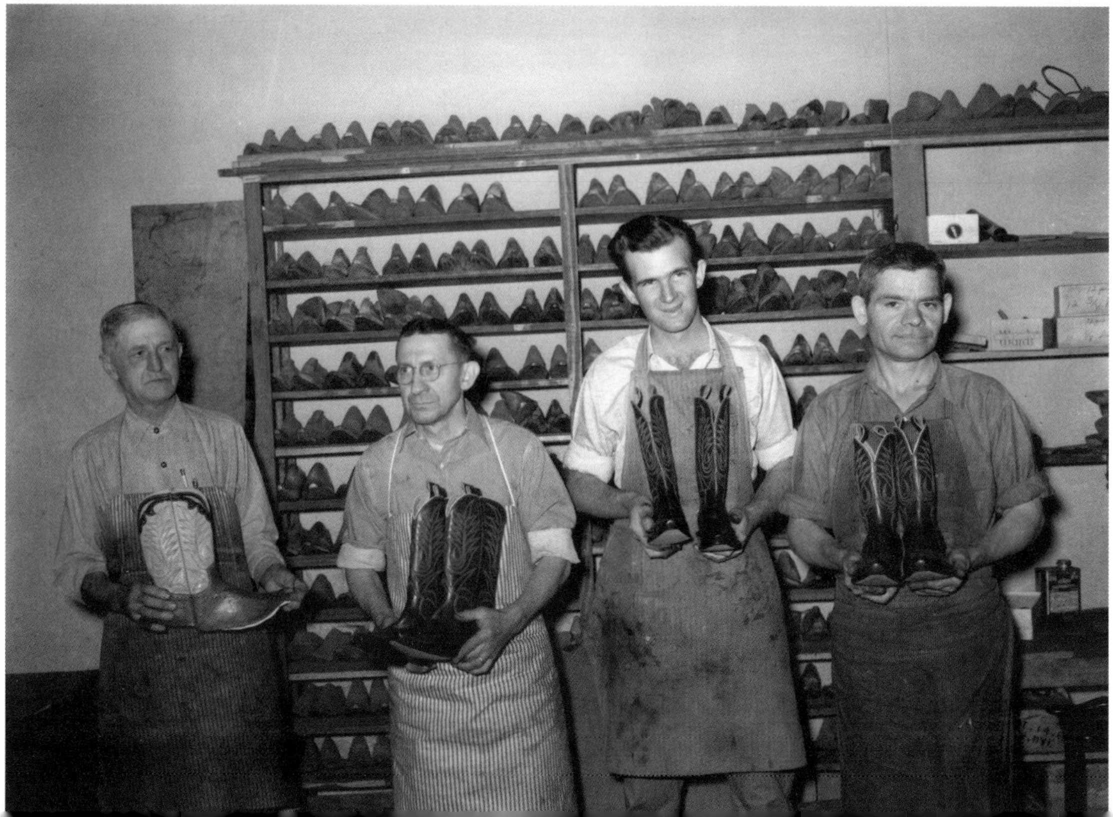

CHUCK WAGONS

Colonel Charles Goodnight is credited with the invention of the chuck wagon in 1866 to feed cowboys on trail drives and roundups. He purchased a war surplus Studebaker wagon, outfitted it with a kitchen on the back, and developed an efficient layout that was soon adopted by trail drovers across the West. The chuck wagon was used to transport food and supplies for trail crews and cattle drives and served as the cowboy cook's home on long trail drives. The chuck wagon also carried provisions for the trail hands, especially water, plus bedrolls, shovels, axes, rope, and even a few personal items. The chuck wagon was sometimes drawn by oxen, but usually by mules. The chuck wagon was the headquarters of every cattle outfit on the range. The cowboys ate their meals there, and it was their social center and recreational spot, a natural gathering place for exchanging "windies," or tall tales, listening to music if there happened to be a musician in the group, or just recounting the experiences of the day. Local makers provided chuck wagons through most of the trail-driving days, but by the 1880s, Studebaker offered its standard "Round-Up Wagon," and other companies such as Springfield Wagon, Old Hickory Wagon, Moline Wagon, and the Mitchell Wagon Company produced chuck wagons ready for the trail. Barbed wire and motor transportation eventually brought the chuck wagon era to a close, although some outfits have continued to use them to this day.

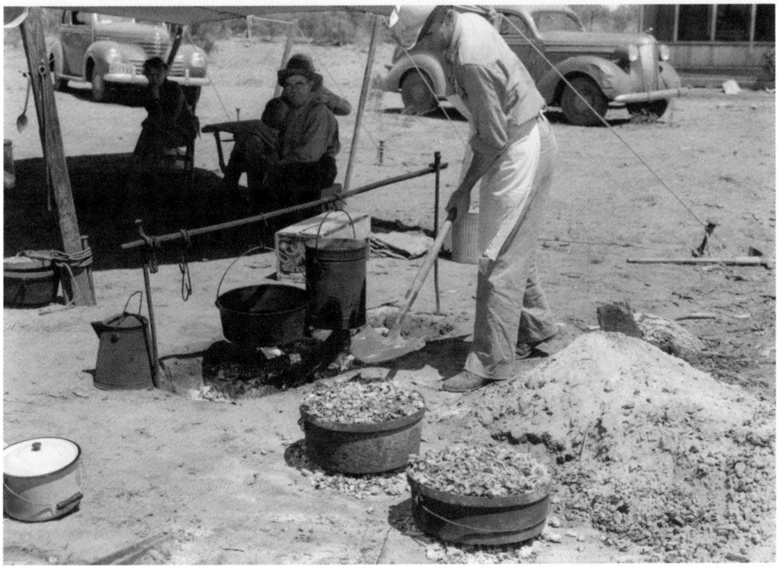

OPPOSITE: Workers in the bootmaking shop. Alpine, Texas. Russell Lee, May 1939. *LC-USF34-033347-D DLC*

LEFT: Cook making bread in front of the chuck wagon. The SMS Ranch near Spur, Texas. Russell Lee, May 1939. *LC-USF34-033323-D DLC*

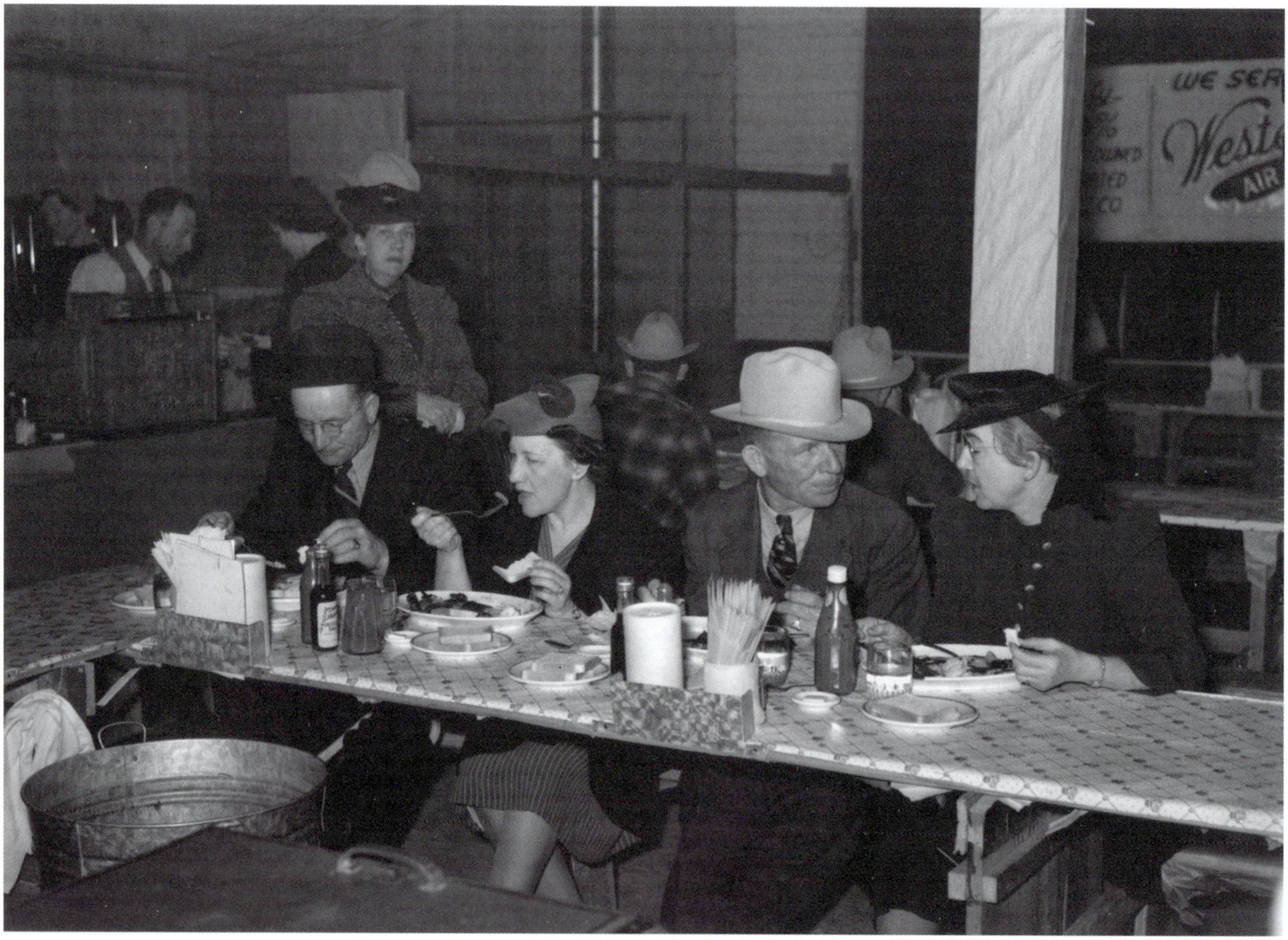

ABOVE: Cattlemen and their wives eating at a restaurant on the grounds of the San Angelo Fat Stock Show. San Angelo, Texas. Russell Lee, March 1940.
LC-USF34-035527-D DLC

OPPOSITE: Saddle shop, Alpine, Texas. Saddle maker tamping leather in place after cementing it. Russell Lee, May 1939.
LC-USF34-033266-D DLC

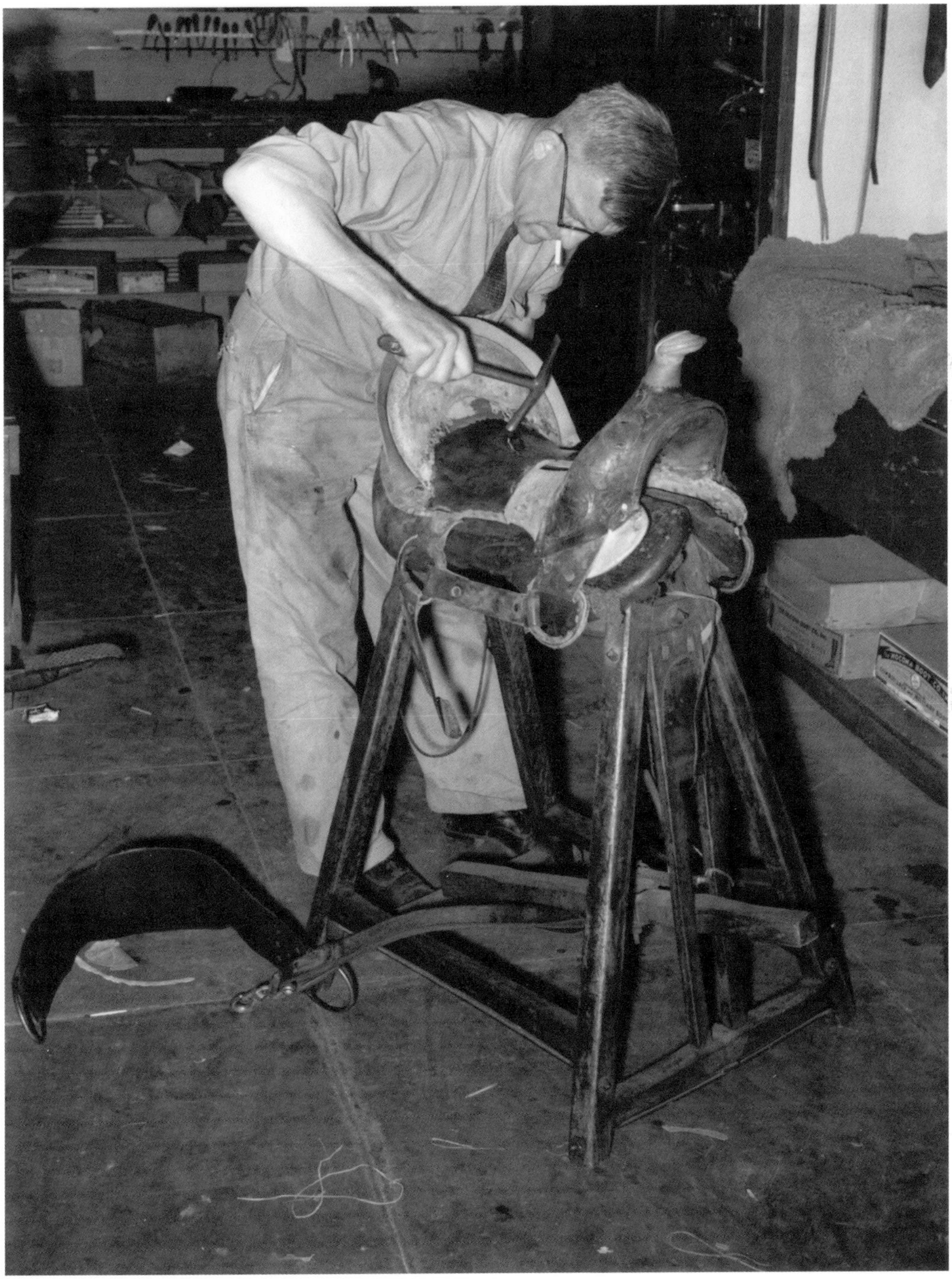

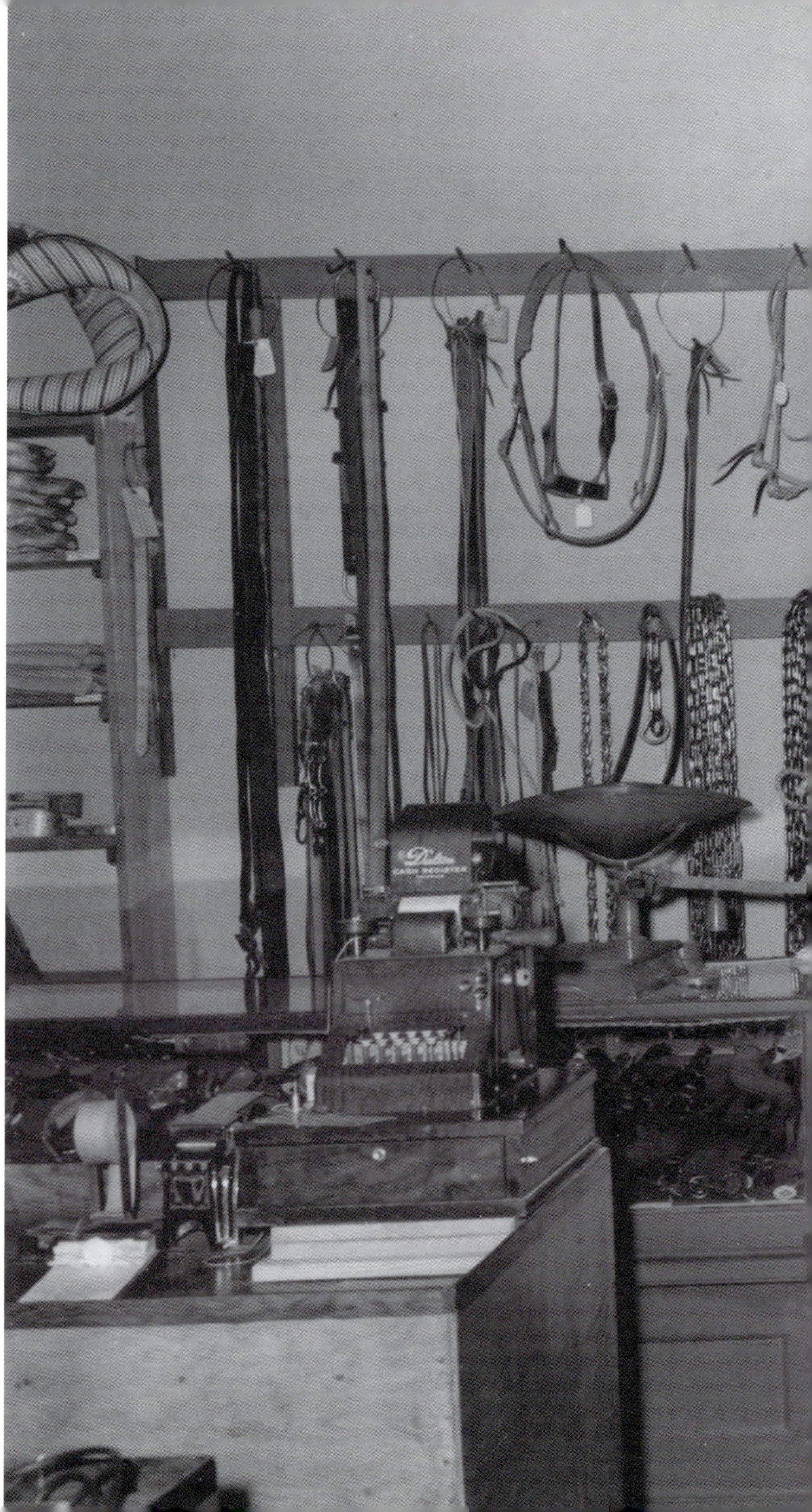

Ranch supply store, Alpine, Texas.
Cowboy examining saddle. Russell Lee,
May 1939.
LC-USF34-033279-D DLC.

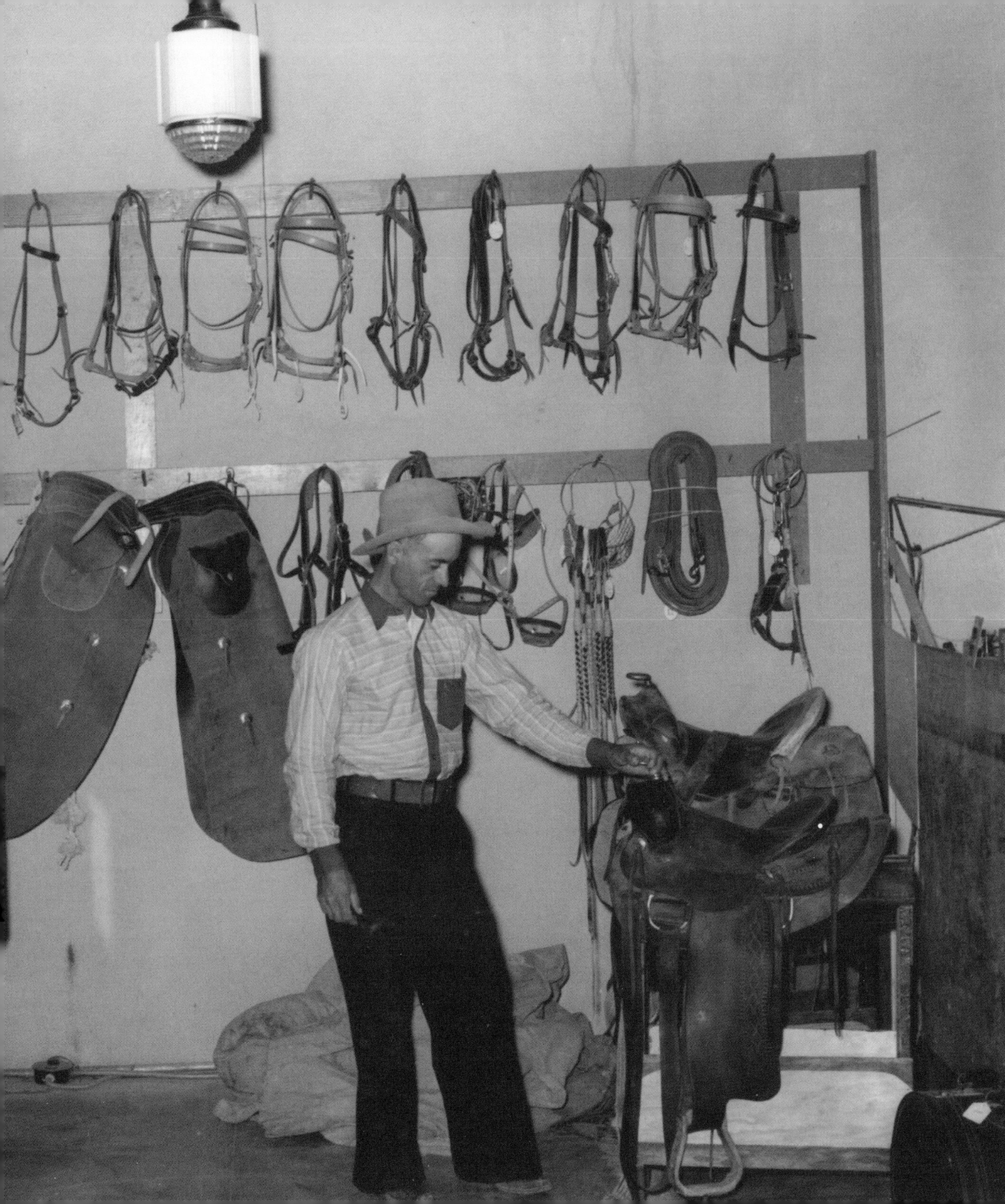

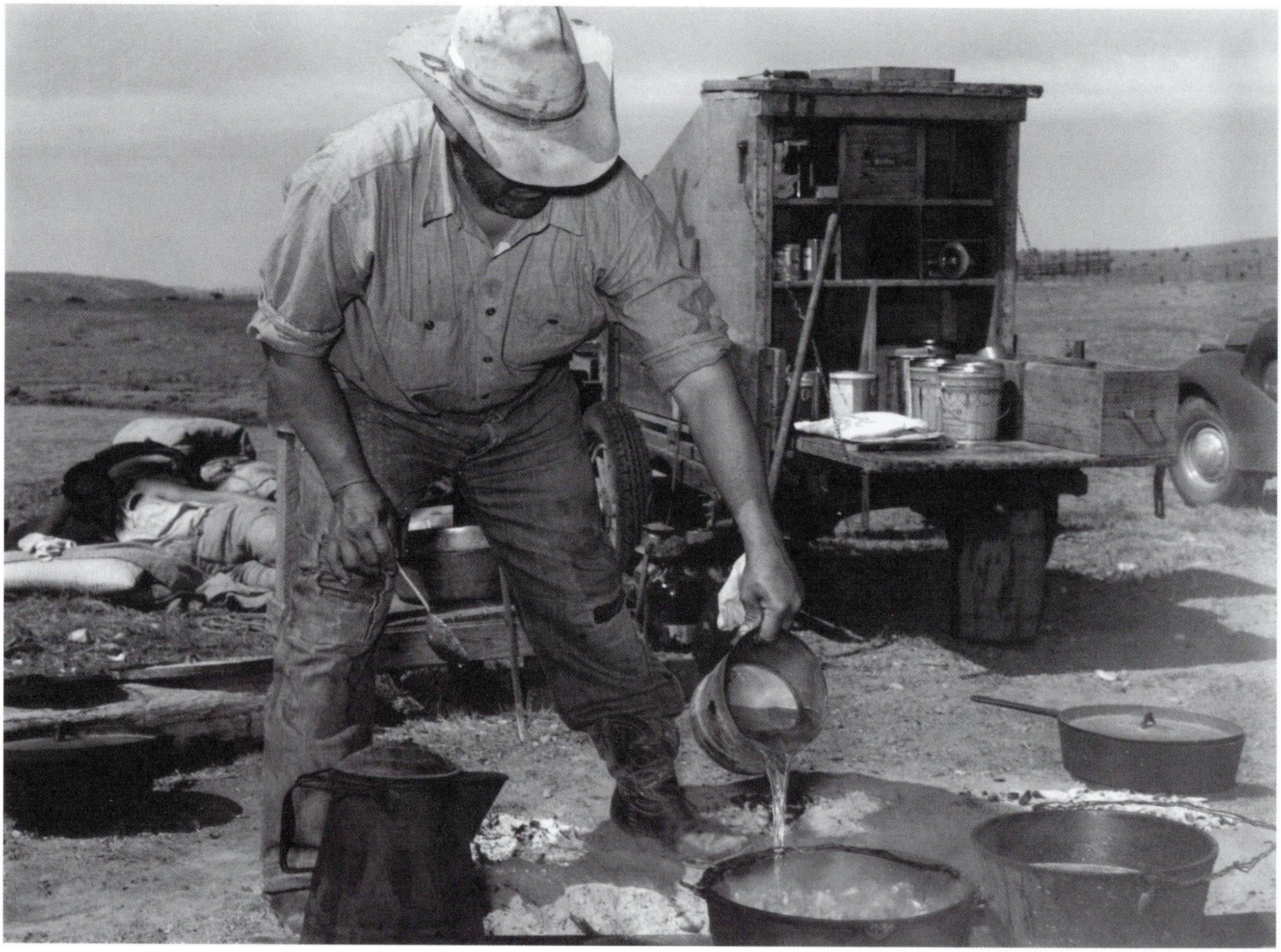

ABOVE: Cattle ranch near Marfa, Texas. Camp cook at work with chuck wagon in background. Russell Lee, May 1939.
LC-USF34-033227-D DLC

OPPOSITE: Ranch in Kimble County, Texas. Shearing a goat with machine equipment. The power is supplied by an automobile that has been converted for this purpose. Russell Lee, March 1940.
LC-USF34-035589-D DLC

SHEEP SHEARING

Before the arrival of electricity and automotive power, sheep shearing was done by hand with scissors that looked similar to grass trimmers. The wool wasn't cut as close to the skin as with electric clippers. Around 1900, companies such as the Chicago Flexible Shaft Company introduced hand-powered mechanical clipping machines. By the early 1930s, shearing clippers powered by electricity—where it was available—and by such means as hooking them up to automobiles created a revolution in the sheep wool industry. An accomplished old-time sheep shearer could shear about 180 sheep a day with hand shears, while a shearer using mechanically or electrically powered shears could shear 300 a day.

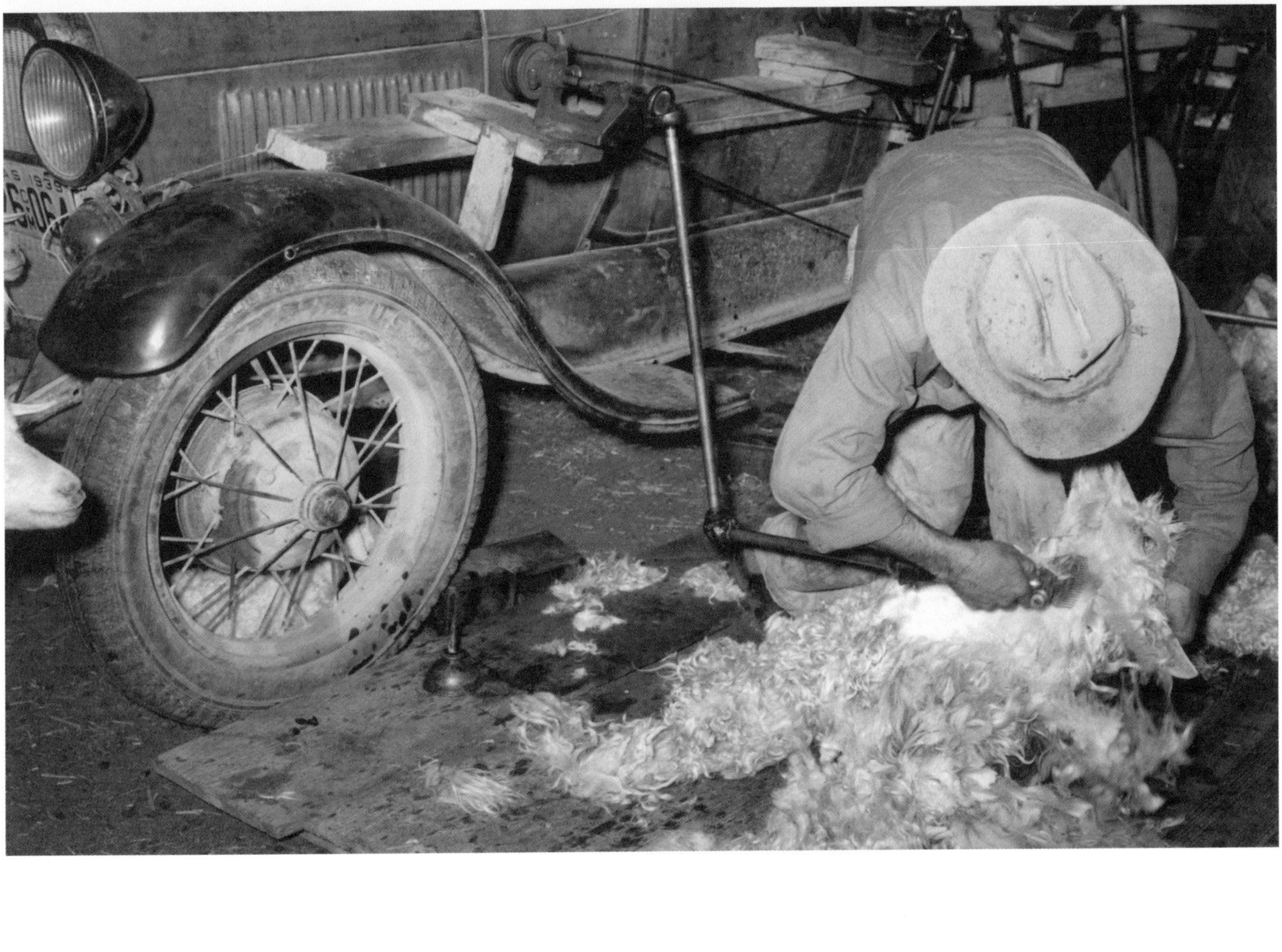

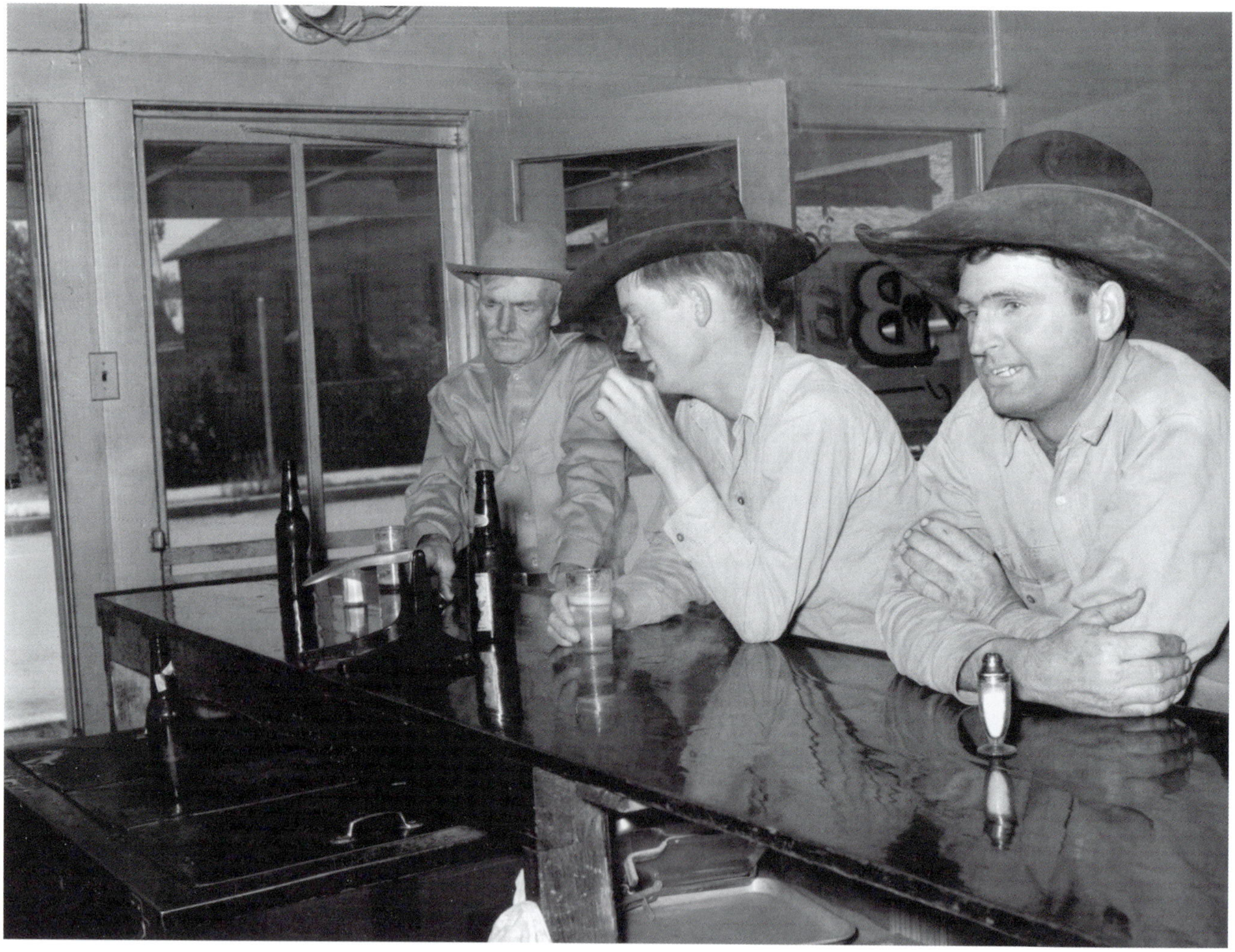

ABOVE: Cowboys in a beer parlor. Alpine, Texas. Russell Lee, May 1939.
LC-USF34-033234-D DLC

OPPOSITE: Old cowboy who was formerly employed at the SMS Ranch near Spur, Texas. Russell Lee, May 1939.
LC-USF34-033325-D DLC

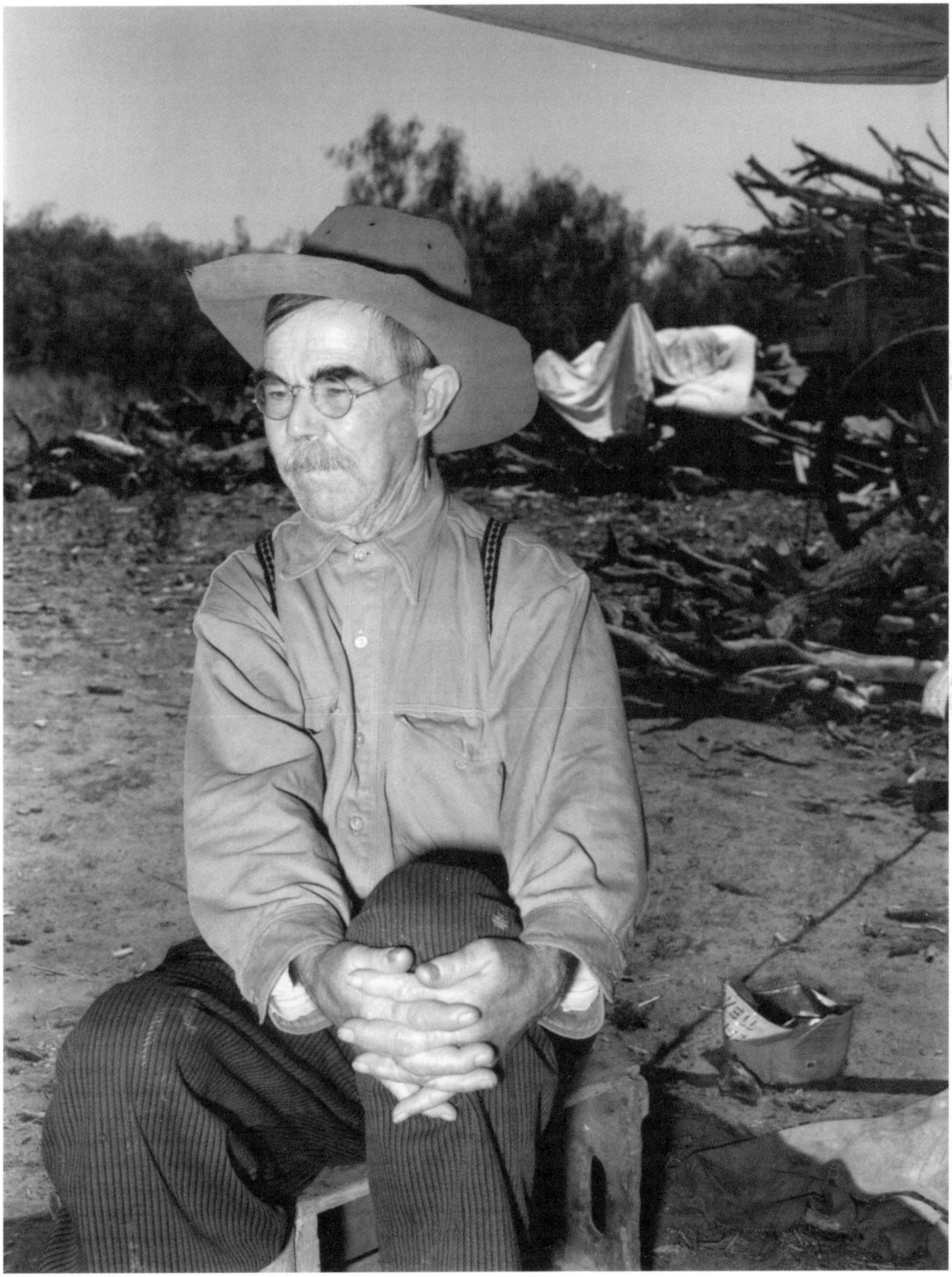

ARRIVAL OF THE AUTOMOBILE

The horse (and mule) reigned supreme on the ranch until the 1920s, when the internal combustion engine brought the automobile onto the scene. Most of the ranching West was definitely "wide open space." Ranches were isolated, with towns and neighbors often miles, and hours, away by horse or buggy. Owning a car allowed ranch folks to go into town more frequently to enjoy the town amenities and culture, perhaps to shop or see a movie. They could also visit neighbors. If someone was injured or ill, they could travel much more quickly for medical help, or the doctor could get to them more quickly. The automobile had a truly transformative effect on ranch life, and everybody wanted one. Many ranches had automobiles even before they had indoor plumbing. One ranch wife, when asked why she wanted a car before indoor plumbing, reportedly replied, "You can't go to town in a bathtub."

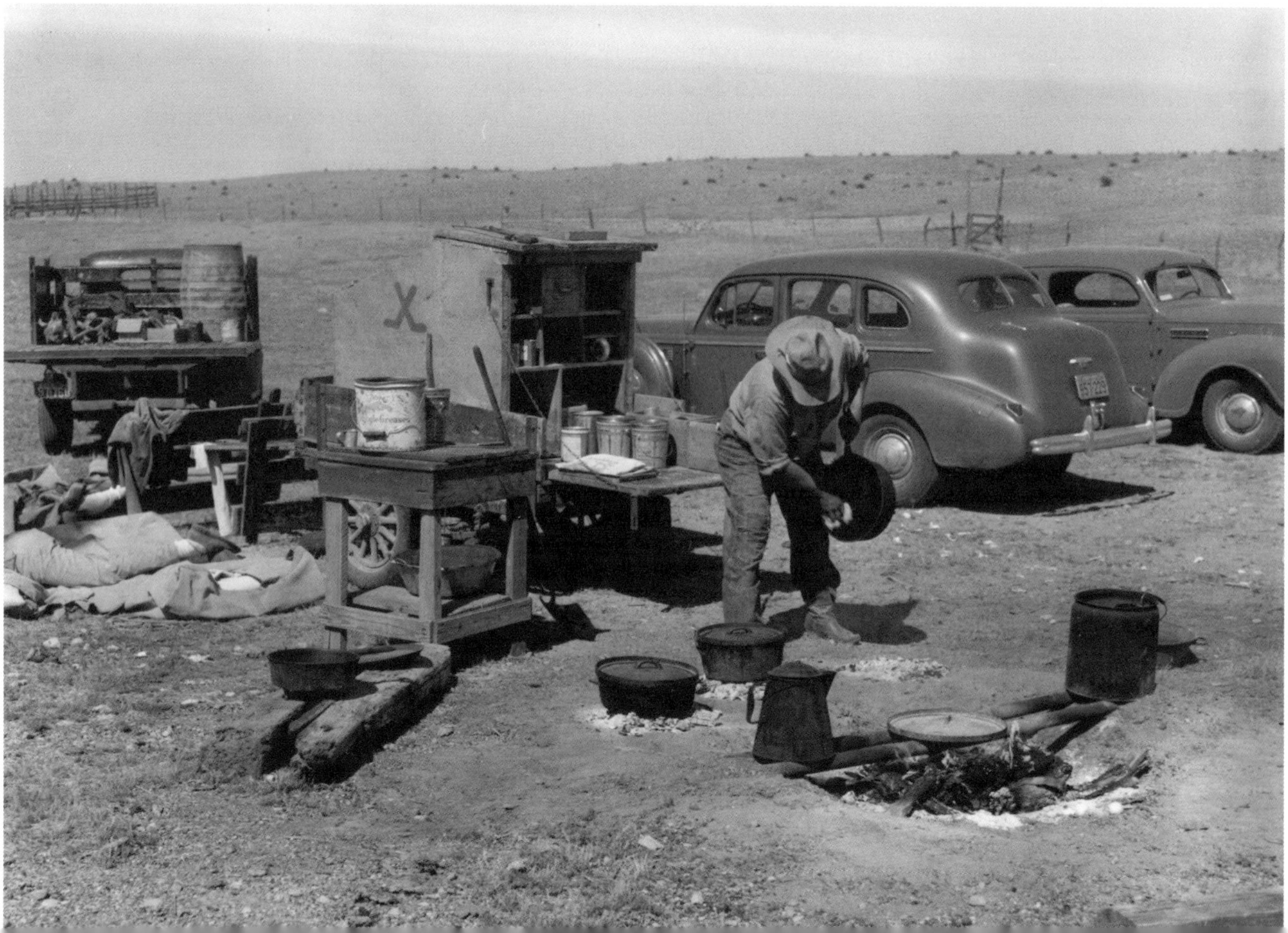

THE BUNKHOUSE

When a cowboy hired on at a ranch, lodging in a bunkhouse was part of the deal, along with meals. As most cowboys were young single men, the standard bunkhouse was a large open room with narrow beds or cots for each individual and with little privacy. One of the best descriptions of a Depression-era bunkhouse is from John Steinbeck in his novella *Of Mice and Men:* "All the walls were whitewashed and the floor unpainted. In three walls there were small, square windows, and in the fourth, a solid door with a wooden latch. Against the wall were eight bunks, five of them made up with blankets and the other three showing their burlap tacking . . . the shelves were loaded with little articles, soap and talcum powder, razors and those magazines ranch men love to read and scoff at and secretly believe." The bunkhouses were often filthy, covered in dust and infested with lice, termites, and rodents. When the cowboys were at the ranch, the bunkhouse was not only where they slept, but also where they socialized, played cards, read, or worked on their gear.

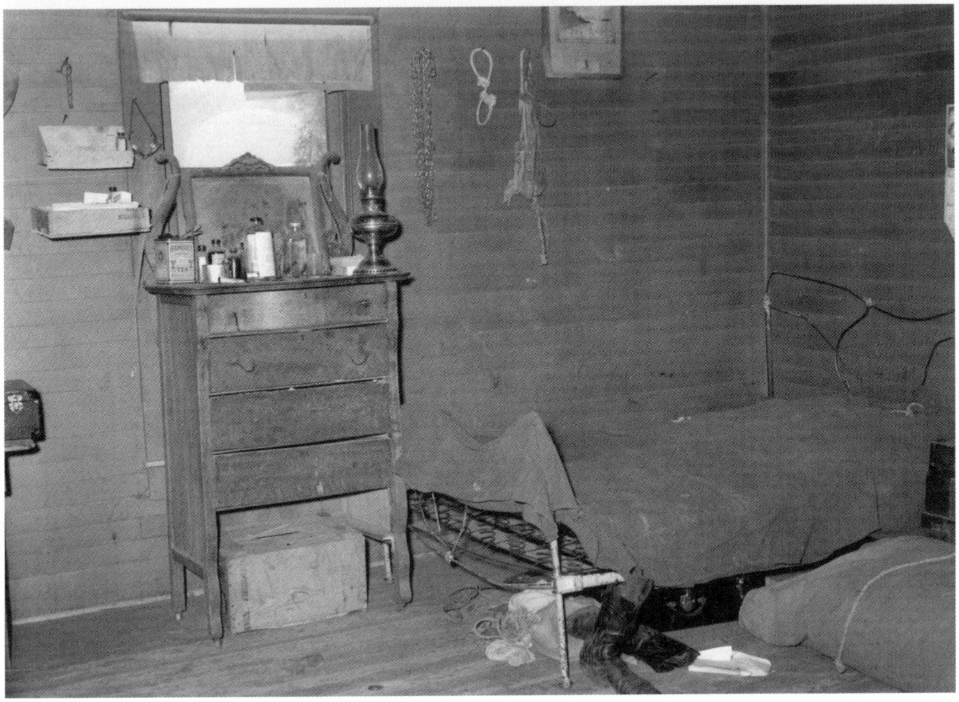

OPPOSITE: Cattle ranch near Marfa, Texas. Camp cook at work with chuck wagon in center and truck for carrying bedrolls at left. Russell Lee, May 1939. *LC-USF34-033247-D DLC*

LEFT: Interior of the bunkhouse at the SMS Ranch near Spur, Texas. Russell Lee, May 1939. *LC-USF34-033361-D DLC*

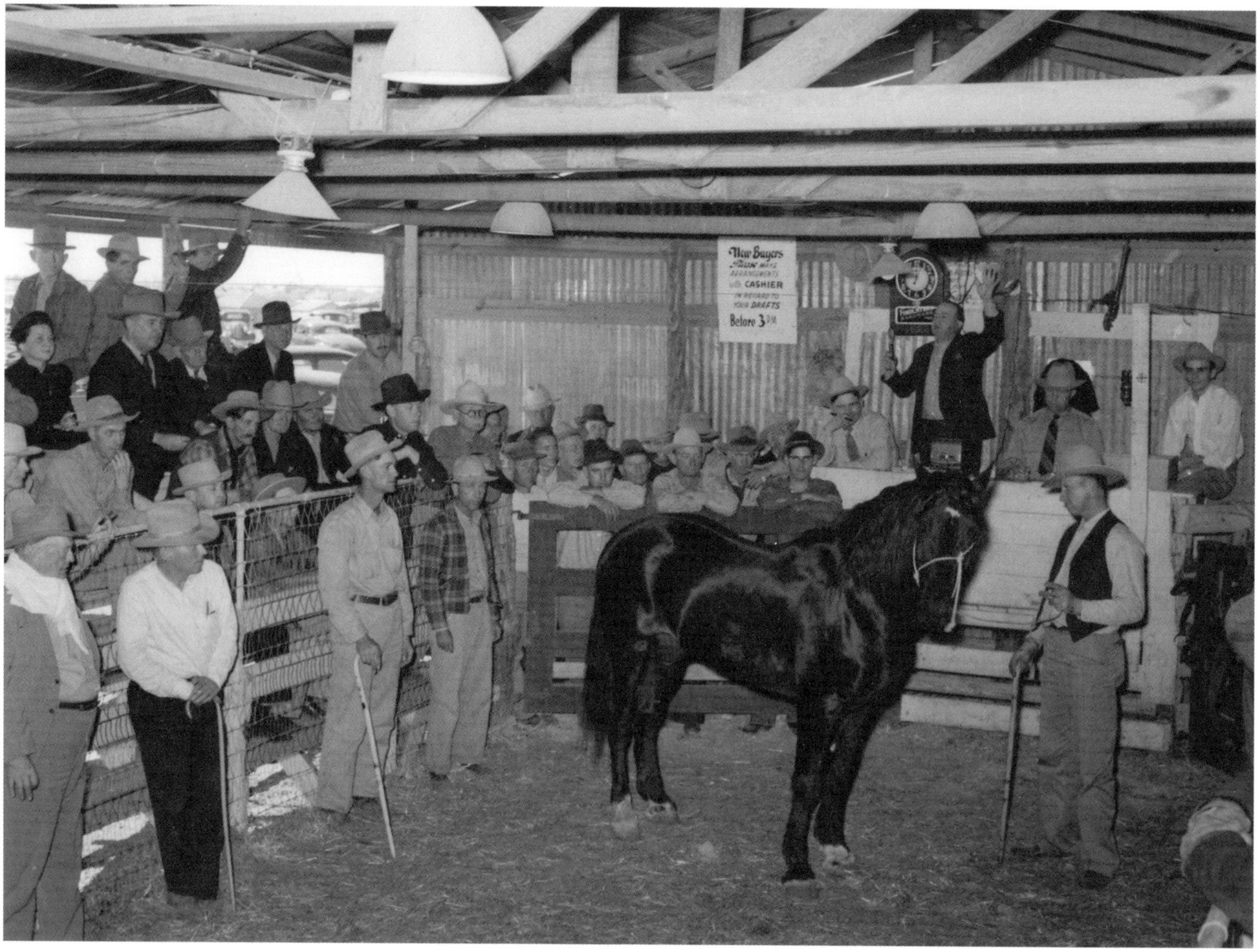
New Buyers
CASHIER
DRAFTS
Before 3

CATTLE TRUCKS

Cattle trucks started rolling off Detroit assembly lines in the 1930s. They brought about a major shift in the way livestock was transported. With the advent of trucking, ranchers could now choose to support their local economy by selling at their local livestock auction market rather than shipping their livestock by rail to distant terminal markets in cowtowns such as Kansas City, Omaha, St. Louis, Denver, and Chicago. The Depression-era federal emphasis on building ranch and farm-to-market roads accelerated this trend. The convenience of trucking, along with the cattlemens' commitment to supporting their local economies, led to a rapid growth in the number of livestock auctions throughout the 1930s.

OPPOSITE: Auction of horses at the San Angelo Fat Stock Show. San Angelo, Texas. Russell Lee, March 1940. *LC-USF34-035479-D DLC*

BELOW: Beef cattle are loaded into this "pullman" truck to be transported to market. Imperial County, California. Russell Lee, May 1942. *LC-USF34-072038-D DLC*

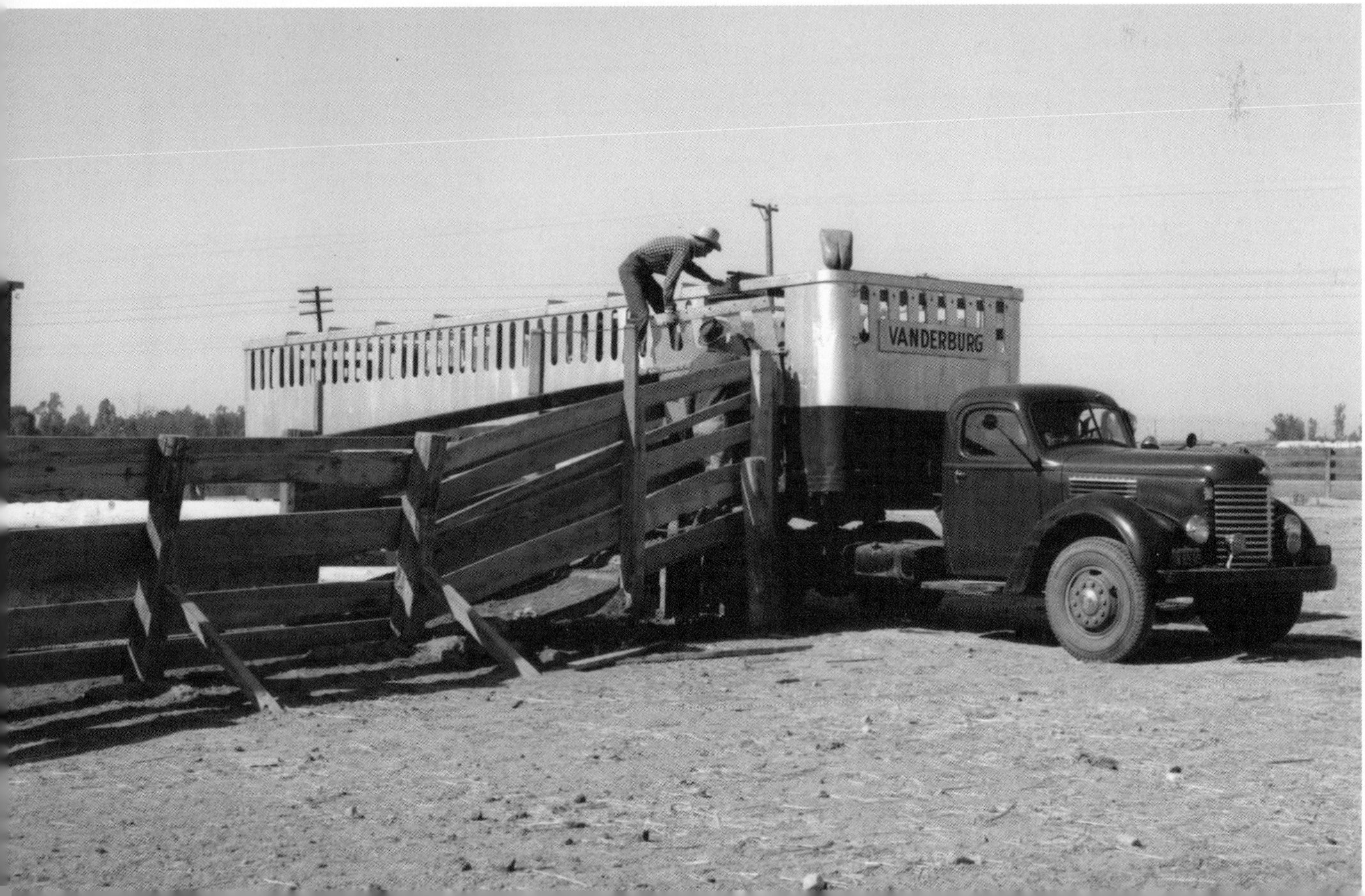

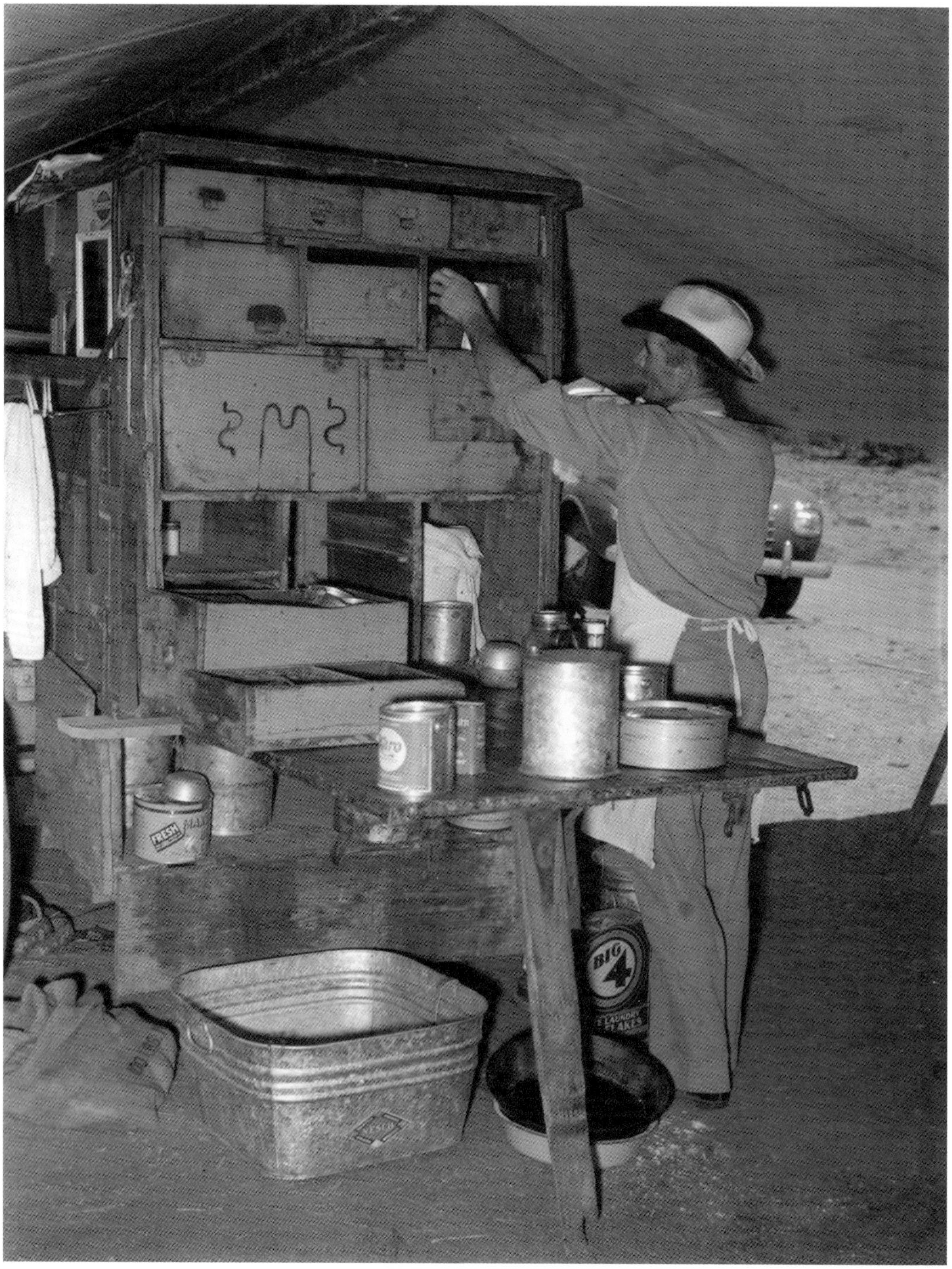
FRESH
BIG 4

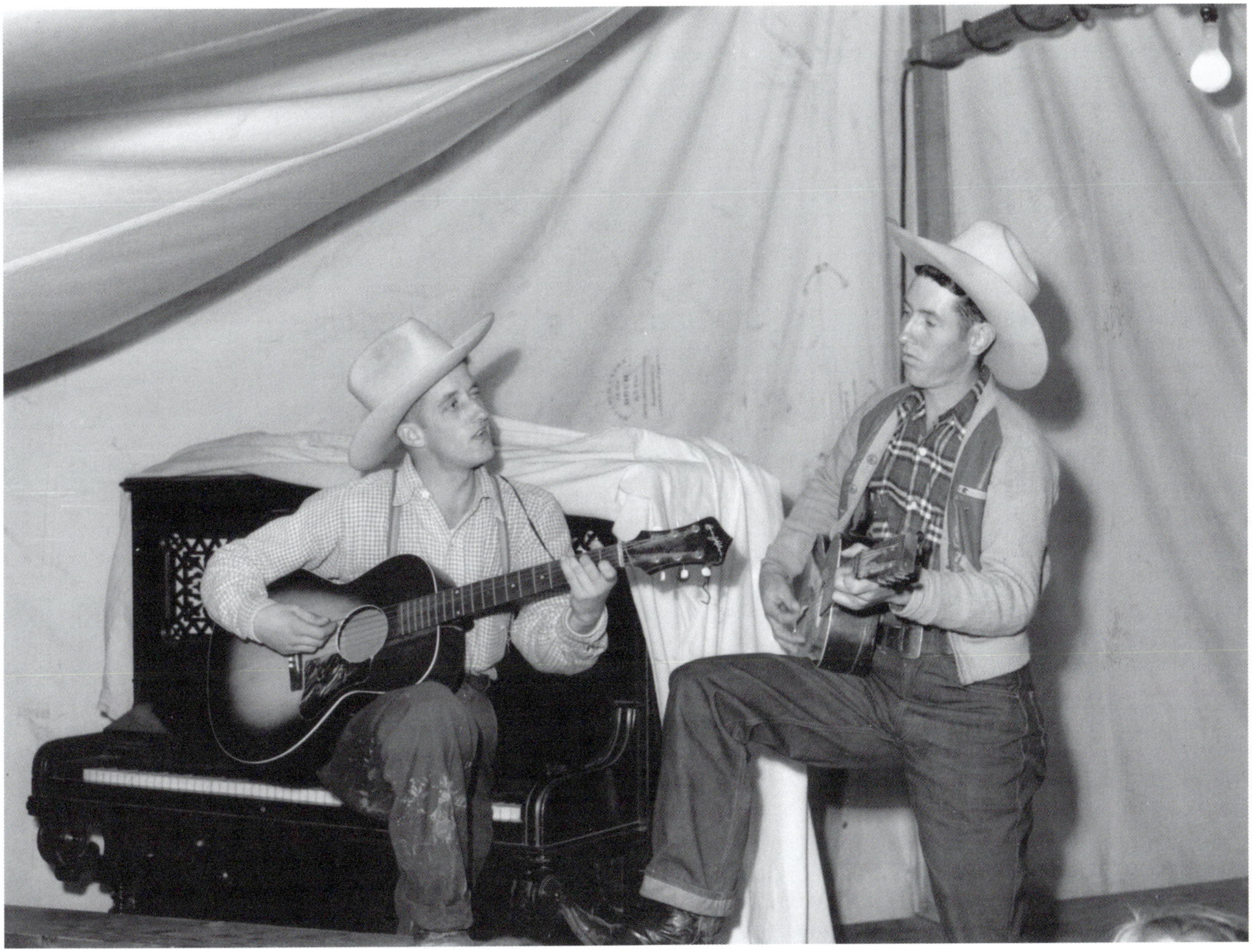

OPPOSITE: Cook at the SMS Ranch in front of the covered end of the chuck wagon. Ranch near Spur, Texas. Russell Lee, May 1939.
LC-USF34-033327-D DLC

ABOVE: Singing cowboy songs for entertainment at the FSA mobile camp for migratory farm workers. Odell, Oregon. Russell Lee, September 1941.
LC-USF34-070358-D DLC

HORSE TRAILERS

Until the advent of the automobile, cowboys had to ride horseback everywhere they went. They rode their horses to where the work needed to be done, often many miles to pastures or a roundup site. With the introduction of automobiles, and particularly the truck, that began to change. Early motor transport on American ranches often consisted of loading a horse into the back of a truck or pickup, usually by backing up to a ramp or an irrigation ditch bank. It didn't take long, however, for people to improvise homemade trailers using car axles and wheels, pulling them with either a truck or a car. Many of these contraptions were, to say the least, not very safe. However, this allowed cowboys to drive to work sites rather than ride, saving many hours in the saddle. It also made travel between ranches and to rodeos much easier. The commercial manufacture of horse trailers as we know them today began in the late 1950s by companies such as the Stidham trailer company in Chickasha, Oklahoma.

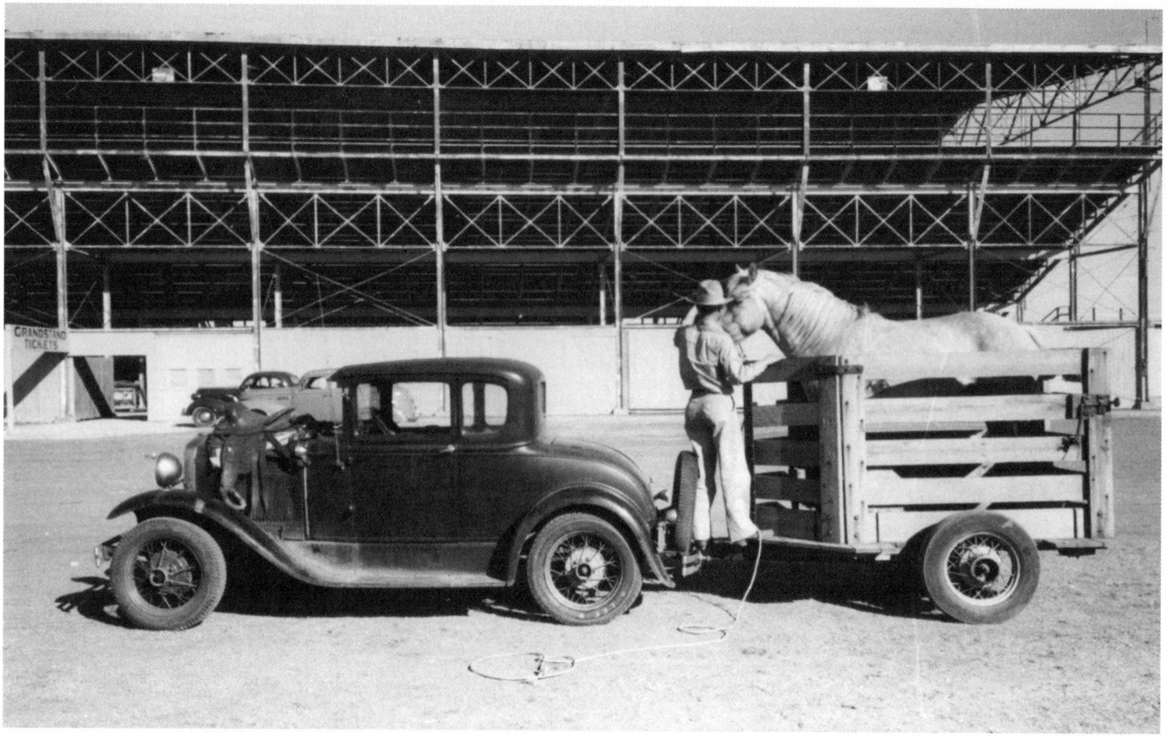

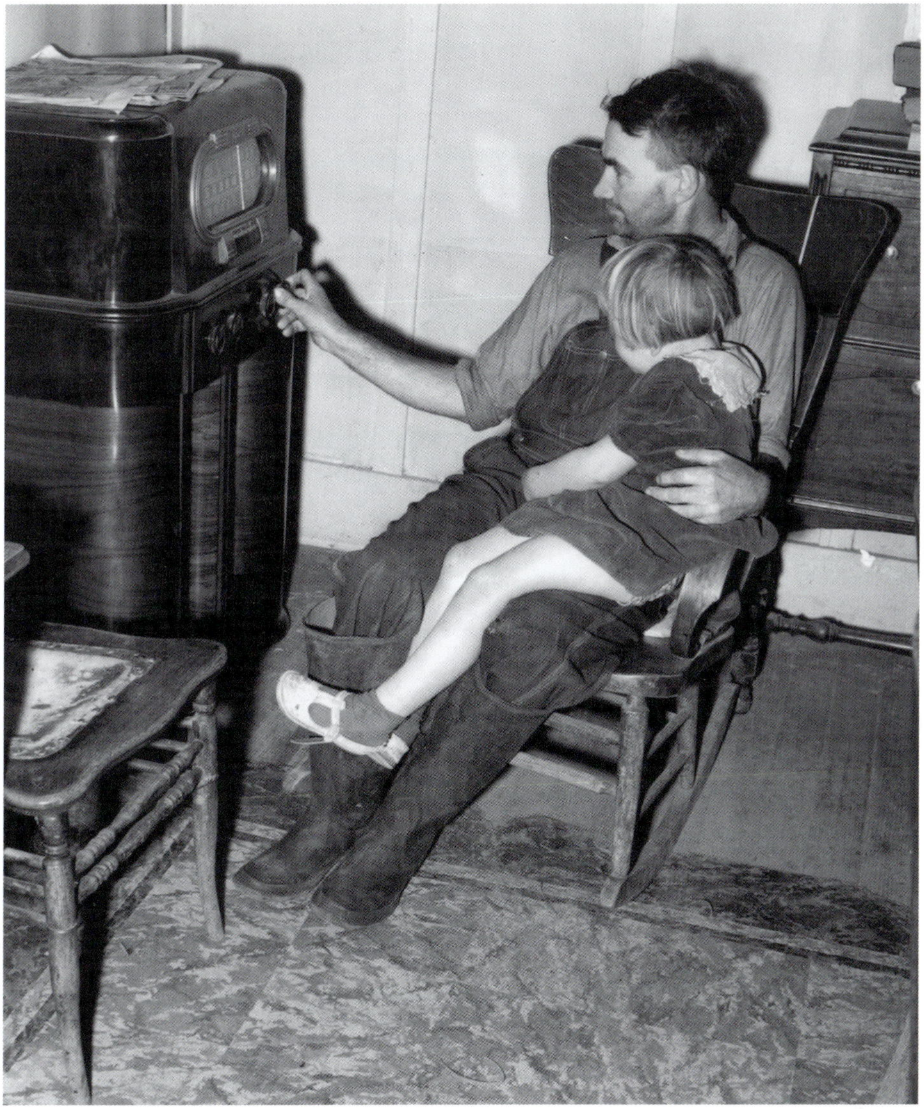

OPPOSITE: Rodeo performer arriving at the San Angelo Fat Stock Show barn with his horse in trailer. San Angelo, Texas. Russell Lee, March 1940.
LC-USF33-012618-M3

ABOVE: John Frost and daughter listening to the radio in their home. Tehama County, California. Russell Lee, November 1940.
LC-DIG-fsa-8b00054 DLC

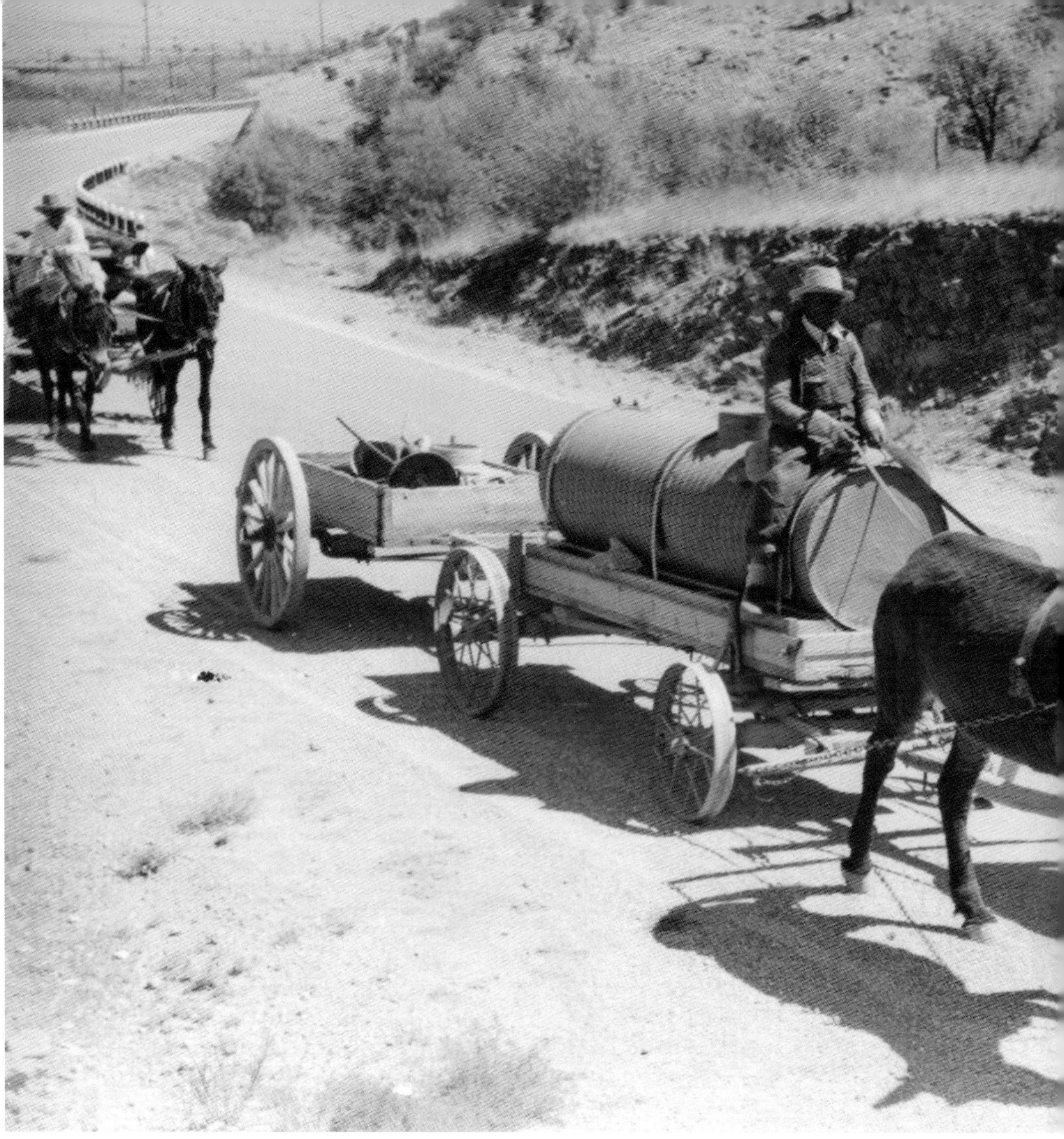

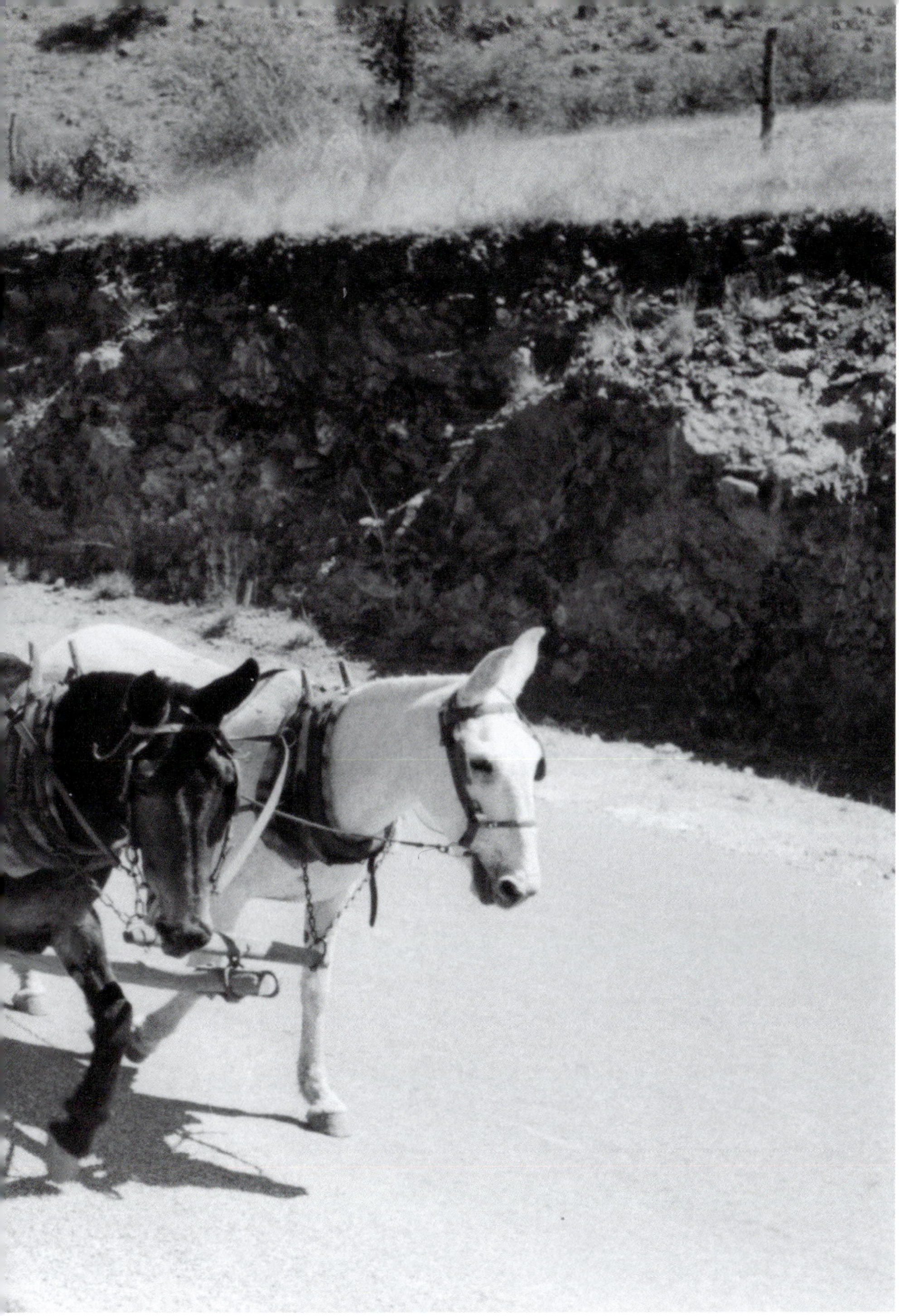

Tank wagon of a cattle ranch on a highway near Marfa, Texas. Russell Lee, May 1939.
LC-USF33-012206-M2 DLC

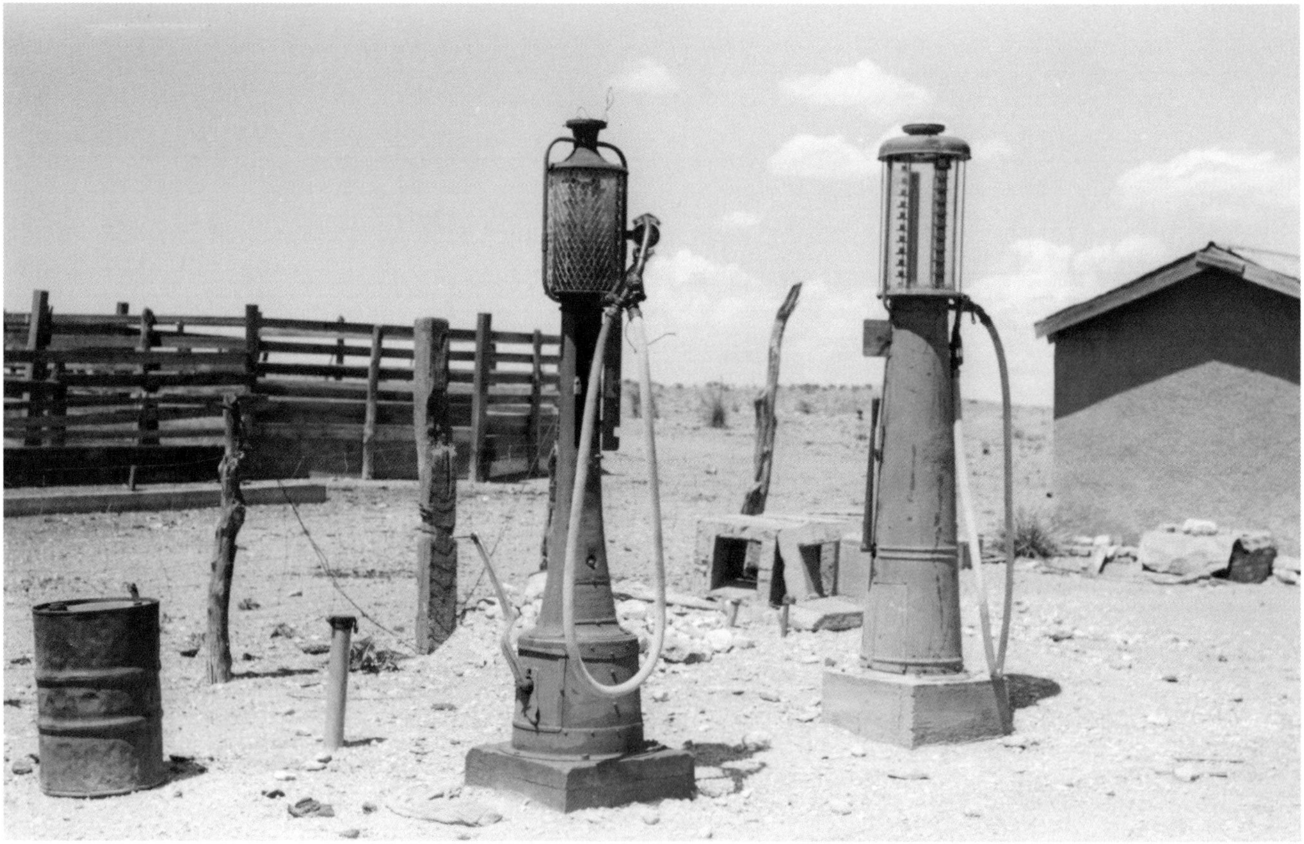

Gasoline pumps on a cattle ranch near Marfa, Texas. Russell Lee, May 1939.
LC-USF33-012208-M4 DLC

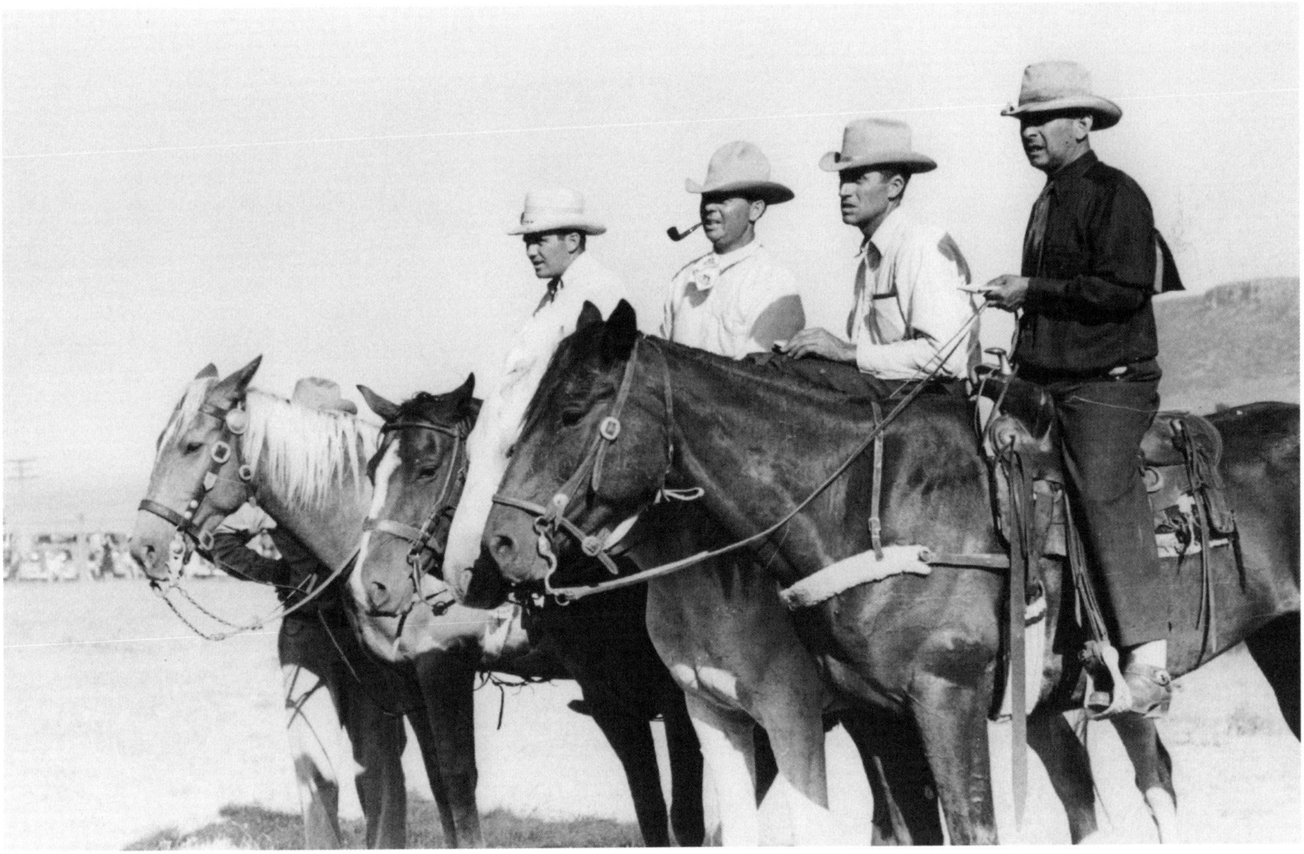

Judges at the Bean Day rodeo. Wagon Mound, New Mexico. Russell Lee, September 1939.
LC-USF3301-012430-M2 DLC

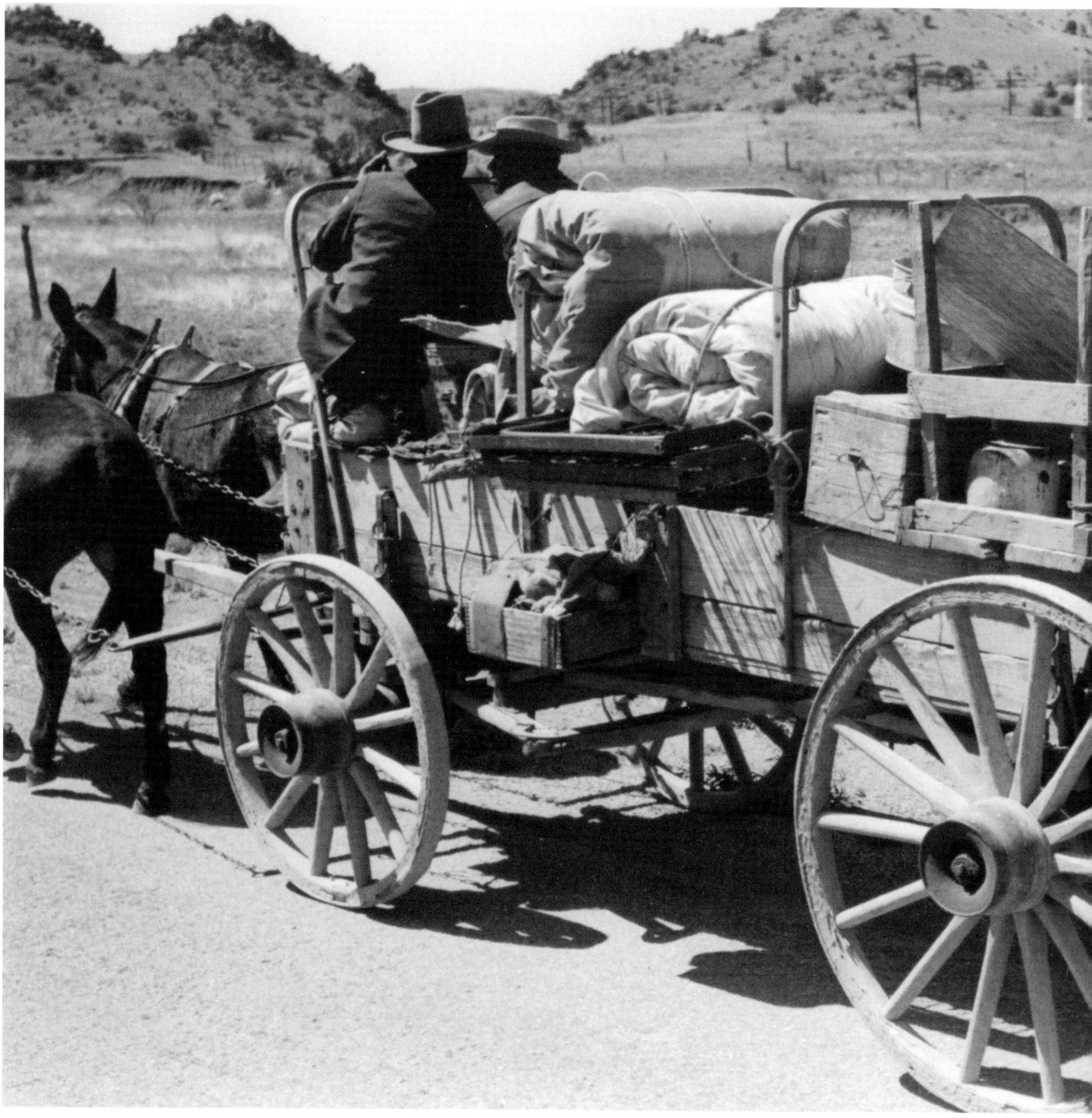

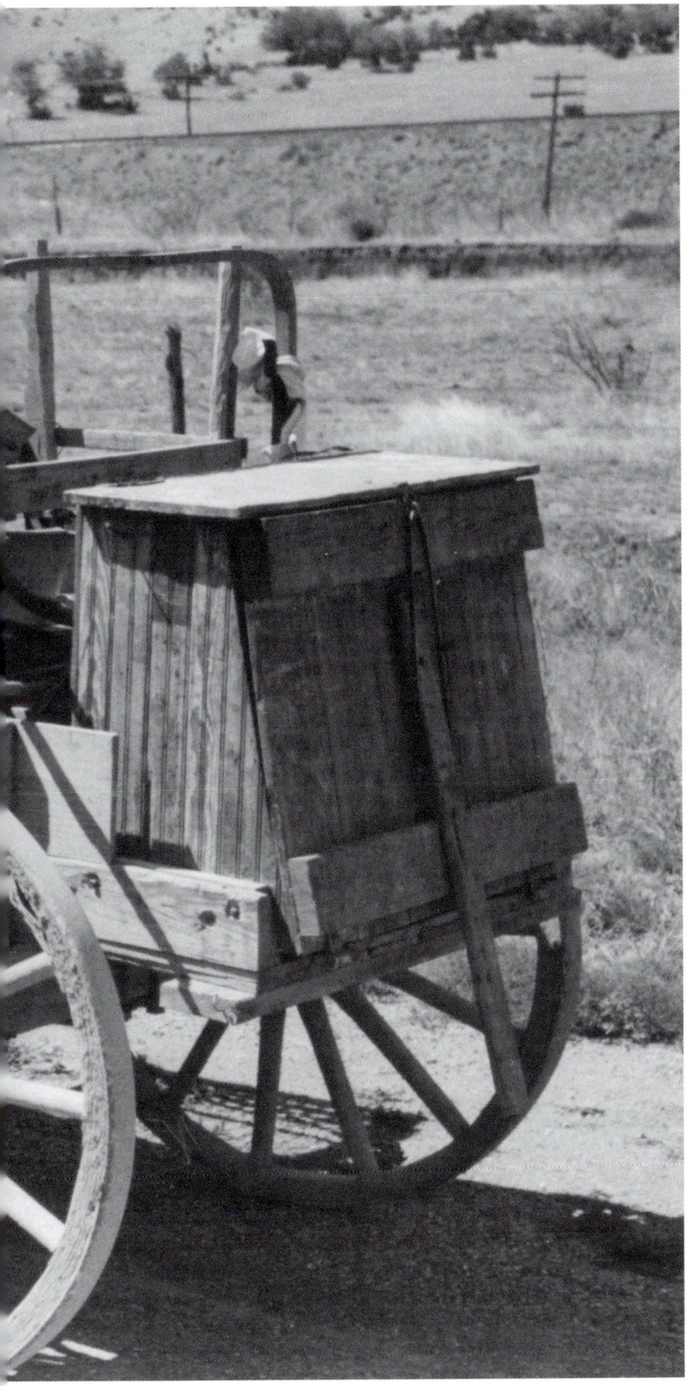

Chuck and bedroll wagon of the tank gang on the highway. Near Marfa, Texas. Russell Lee, May 1939.
LC-USF33-012204-M2 DLC

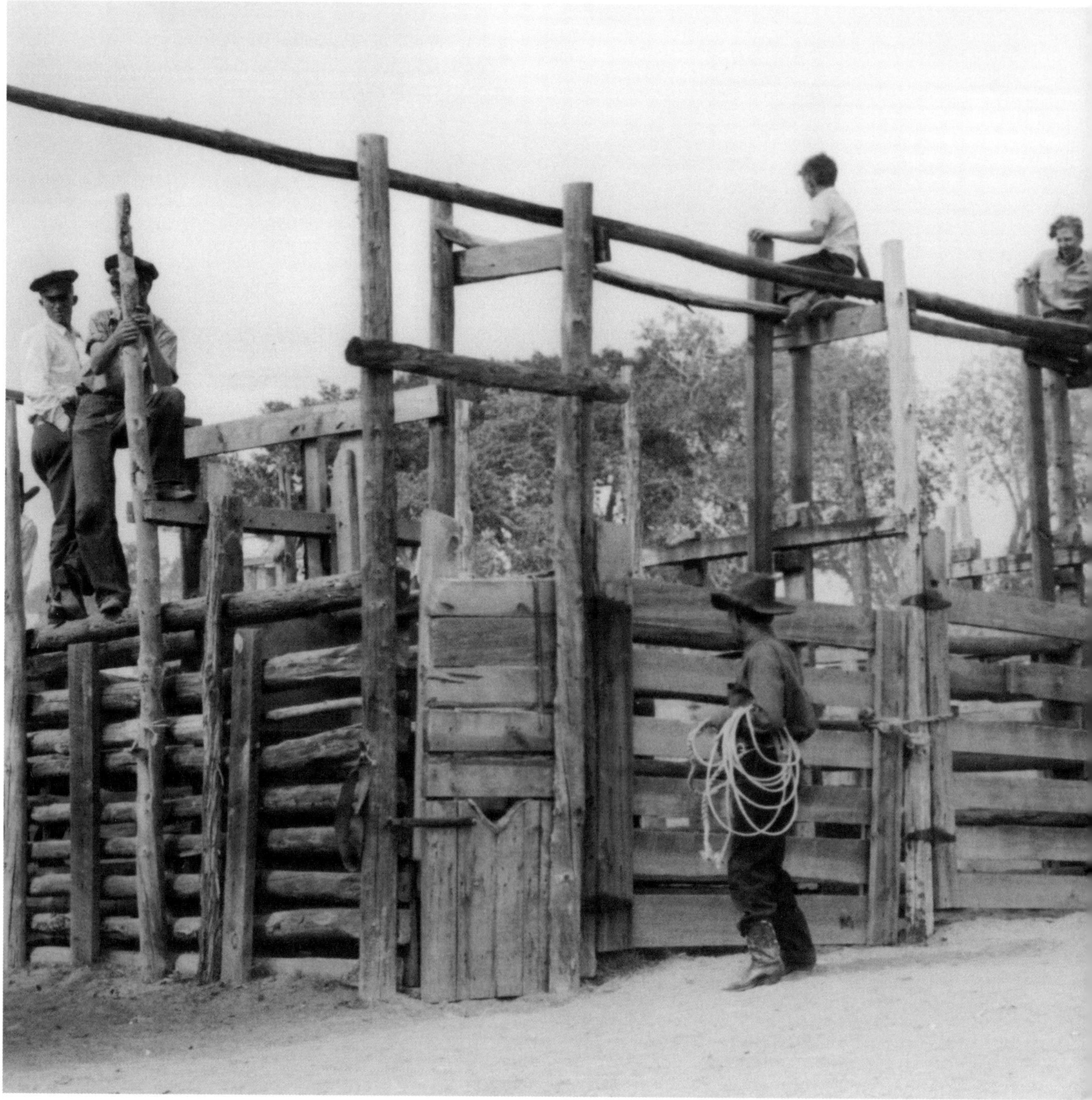

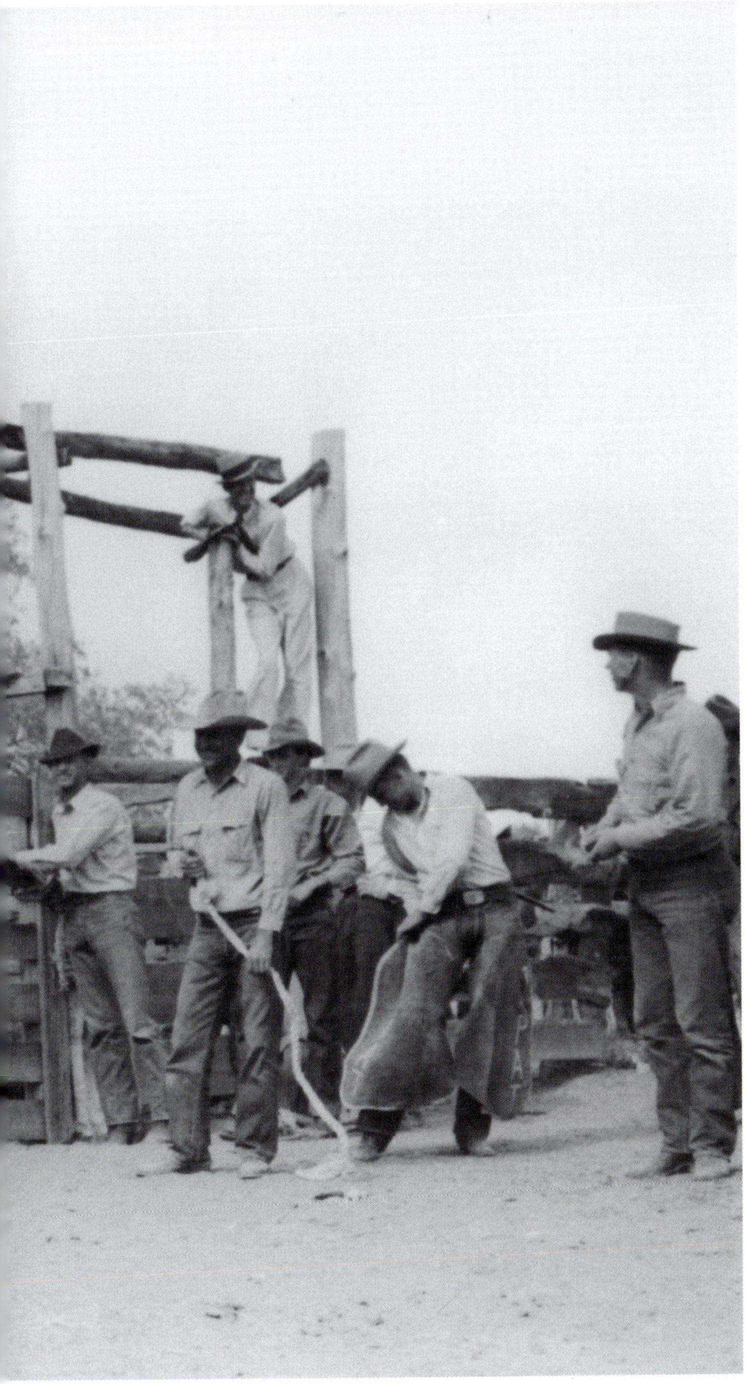

Cowboys around the corral at the rodeo. Quemado, New Mexico. Russell Lee, June 1940.
LC-USF33-012770-M5

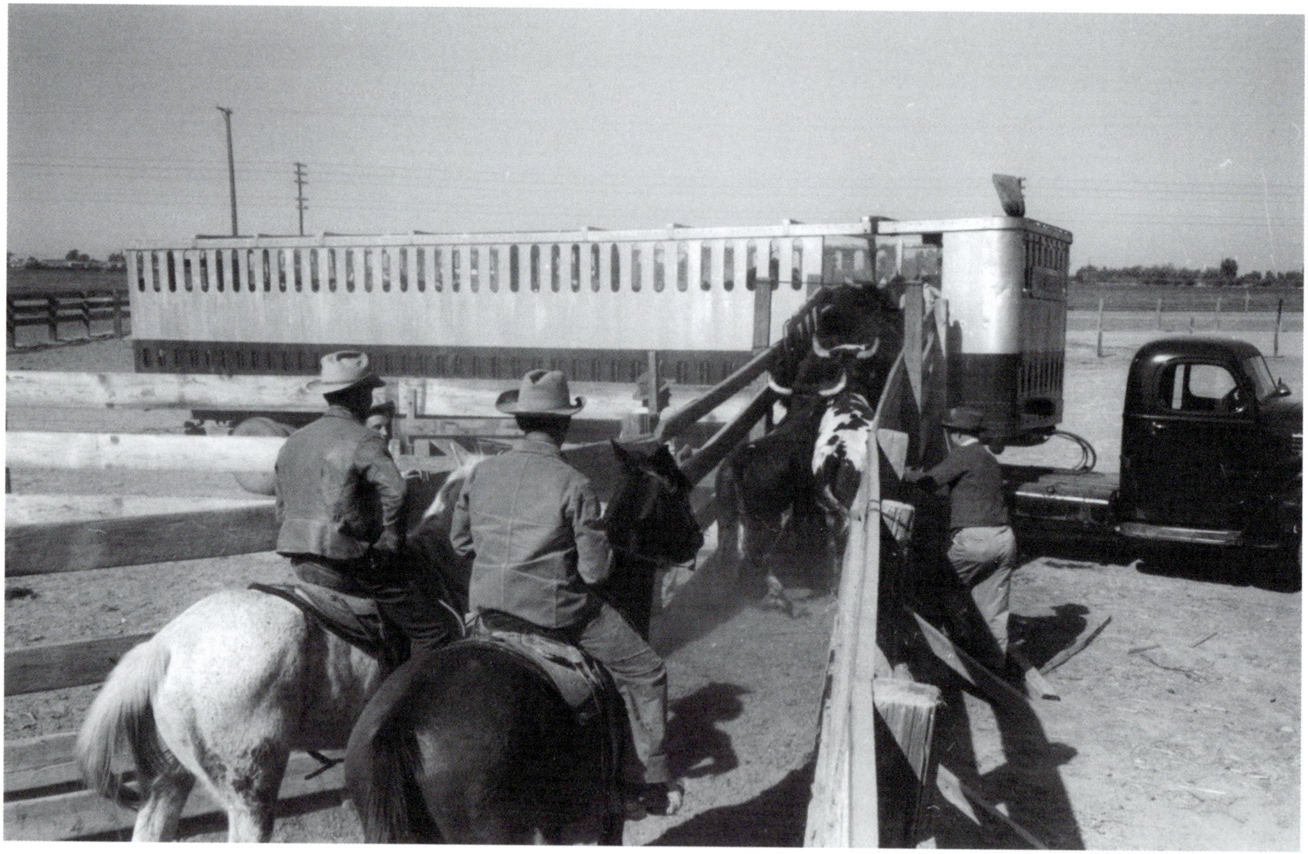

ABOVE: Loading cattle into the trailer for shipment to market. Brawley, California. Russell Lee, March 1942.
LC-USF33-013264-M1 DLC

OPPOSITE: Detail of cowboy on horseback. Ola, Idaho. Russell Lee, July 1942.
LC-USF34-073942-E

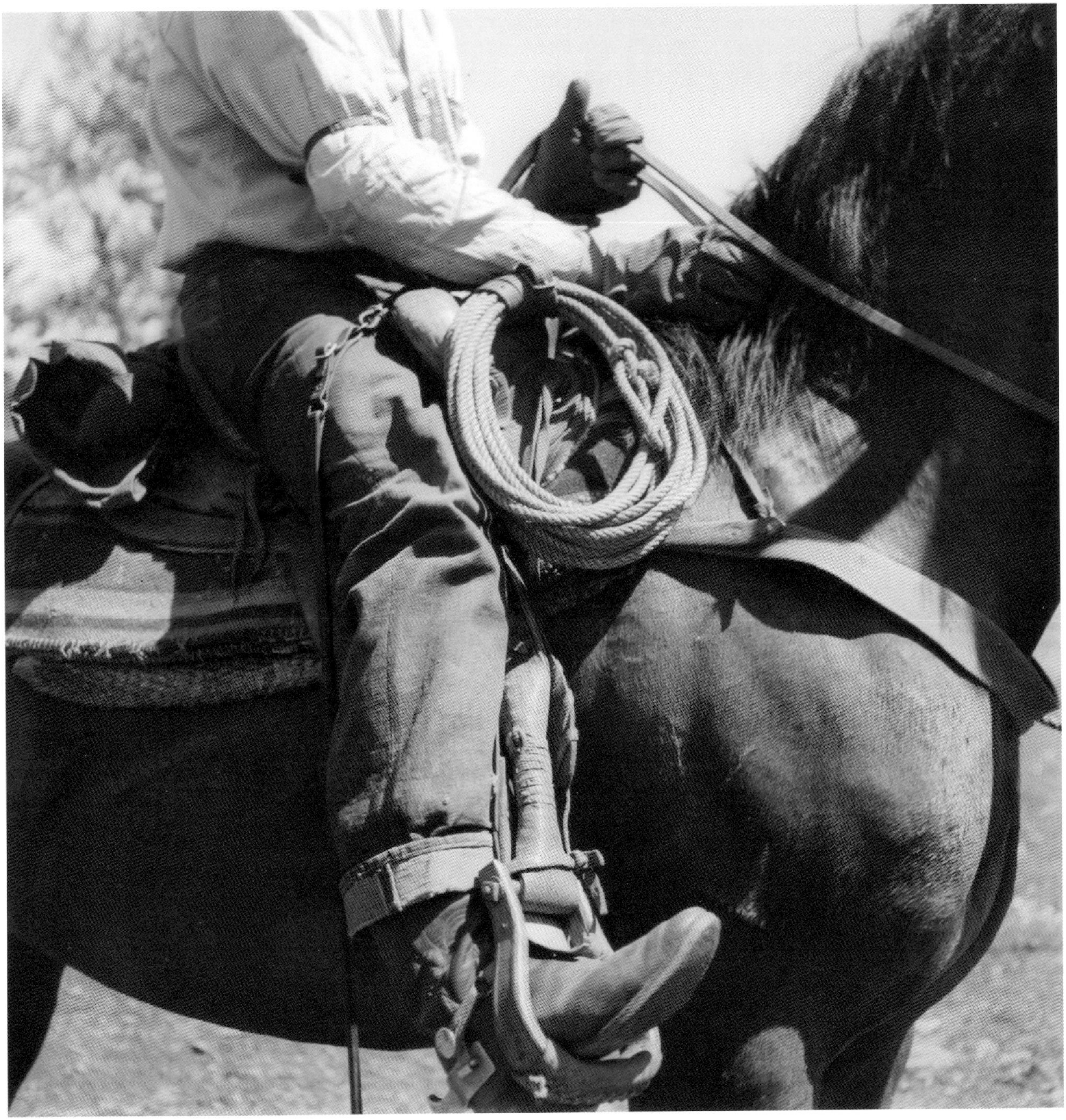

MARION POST WOLCOTT

1910–1990

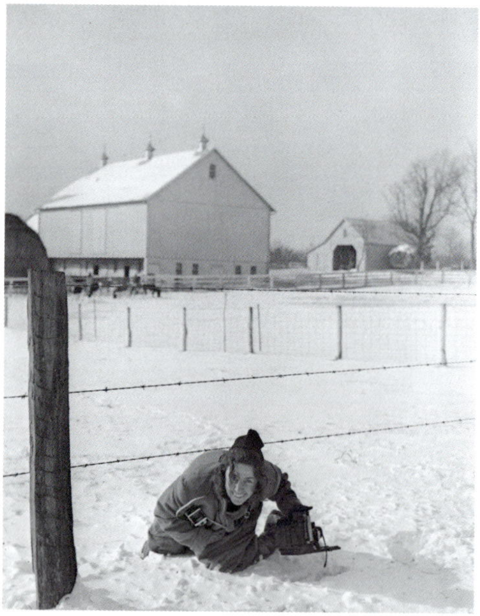

Marion Post Wolcott with Rolleiflex and Speed Graphic in hand in Montgomery County, Maryland. Arthur Rothstein, January 1940.
LC-DIG-ppmsc-00195 DLC

Marion Post was born in New Jersey on June 7, 1910. Post trained as a teacher and went to work in a small town in Massachusetts. When the school closed in the early 1930s, she went to Europe to study at the University of Vienna with her sister Helen. Helen was studying with Trude Fleischmann, a Viennese photographer. Marion Post showed Fleischmann some of her photographs and was encouraged to pursue photography. In 1934 as pre-war tensions mounted in Austria, Post moved to New York. She began freelancing and landed a picture on the cover of the *New York Times Magazine*. She also began attending meetings of the New York Photo League, an important organization that was influencing many of the country's best young photographers. There Marion met Ralph Steiner and Paul Strand, who, upon seeing her work, asked her to join a group of serious young photographers who met at Steiner's apartment to discuss and critique each other's photography. She later became a staff photographer for the *Philadelphia Evening Bulletin*. Her friend Steiner introduced her to Roy Stryker, who hired her in 1938, sending her to photograph West Virginia coal miners as her first assignment.

In June 1941, she married Lee Wolcott, an assistant to Henry Wallace, Secretary of Agriculture under President Roosevelt. In August and September, she found herself in the Rocky Mountains, photographing in Wyoming and Montana. Shortly after that, in 1942, she left the FSA and devoted herself to raising her children and working a Virginia farm with her husband. She took relatively few photographs until the 1970s, when a renewed interest in Wolcott's images among scholars rekindled her own interest in photography. In 1978, she mounted her first solo exhibition in California, and by the 1980s, the Smithsonian and the Metropolitan Museum of Art began to collect her photographs.

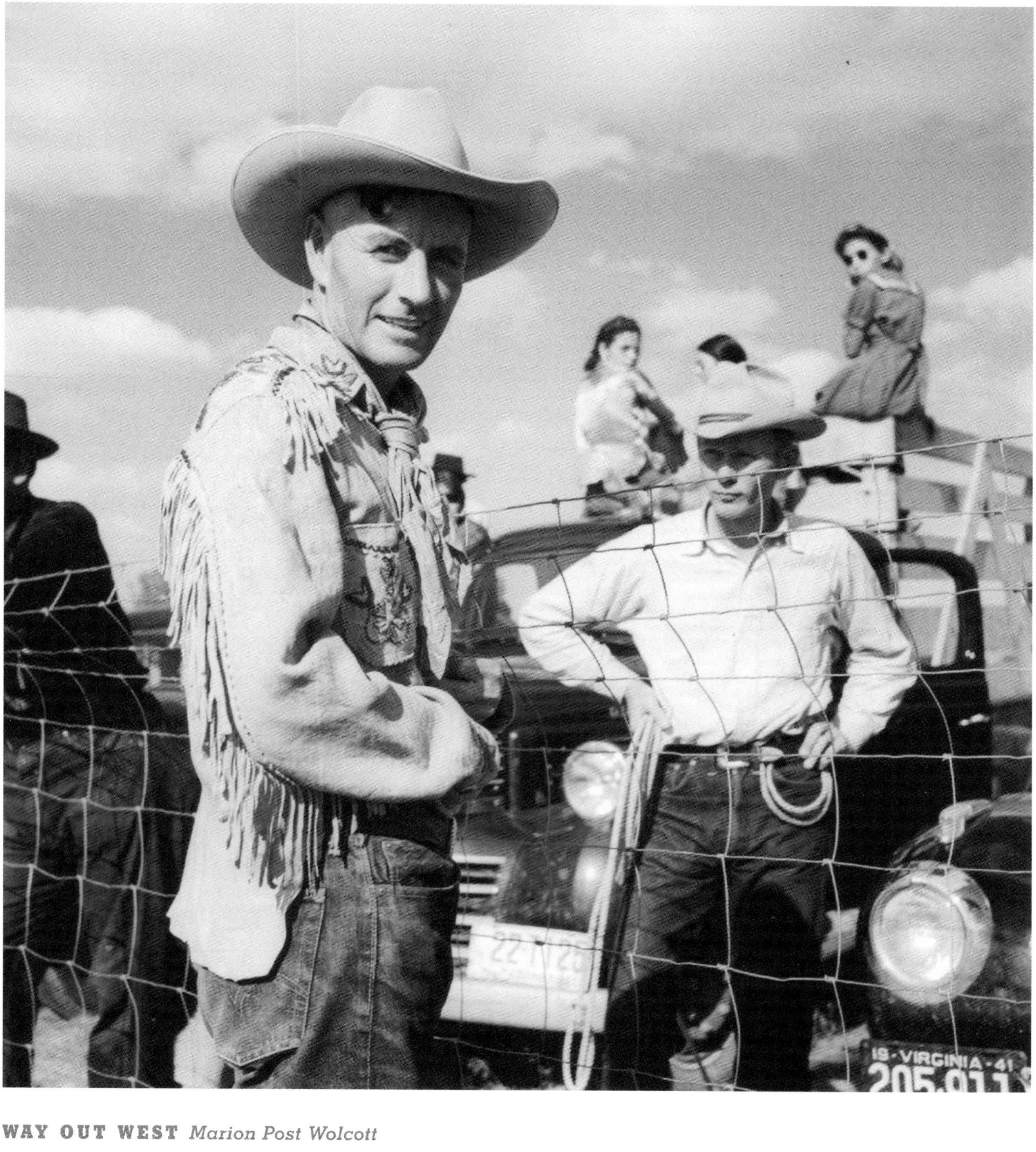
19-VIRGINIA-41

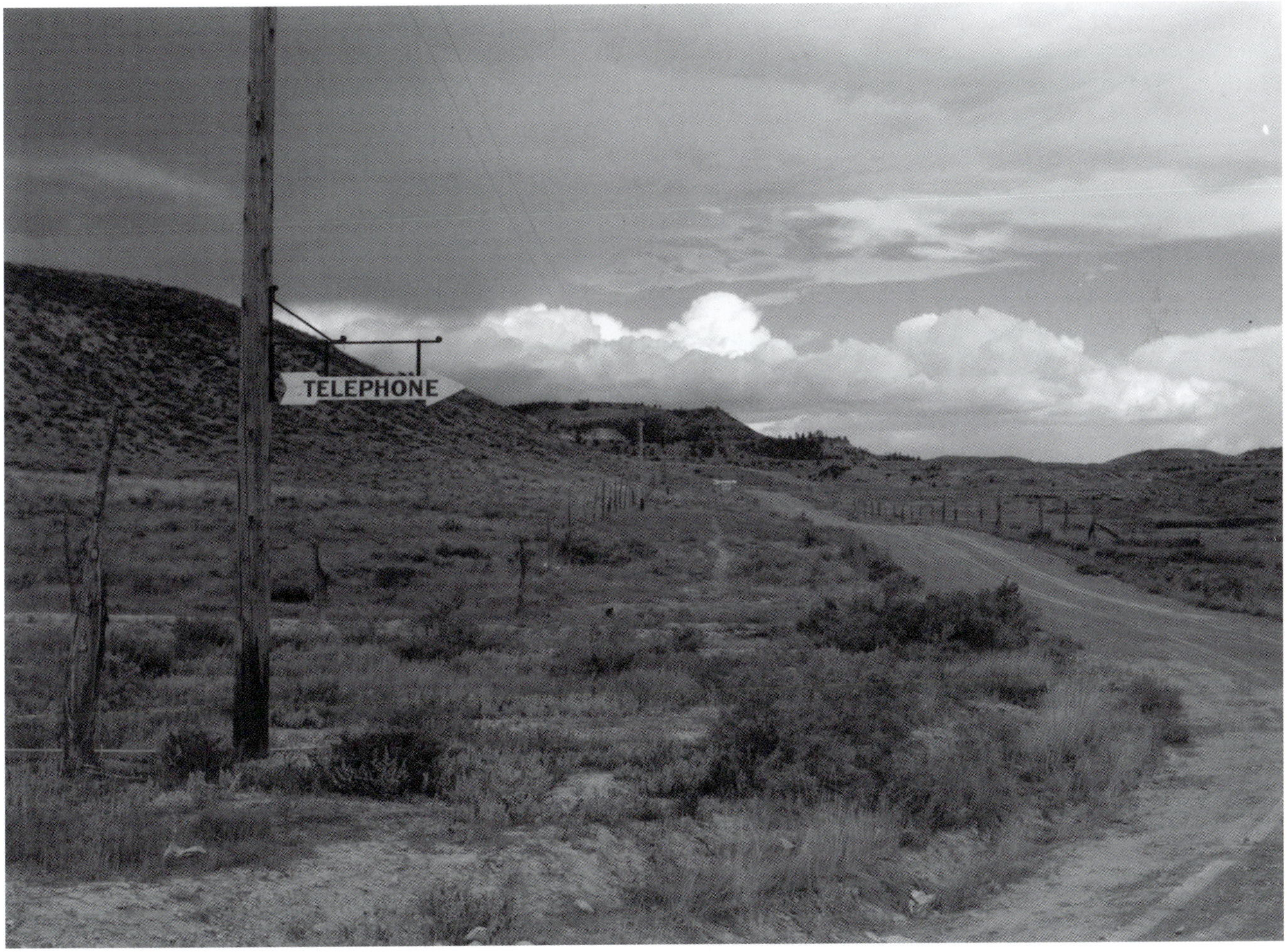

OPPOSITE: Lyman Brewster of Quarter Circle U Ranch at the rodeo in Ashland, Montana. Marion Post Wolcott, September 1941.
LC-USF34-058890-E DLC

ABOVE: Sign at crossroads pointing the direction to the nearest telephone in ranch country. Birney, Montana. Marion Post Wolcott, August 1941.
LC-USF34-058514-D DLC

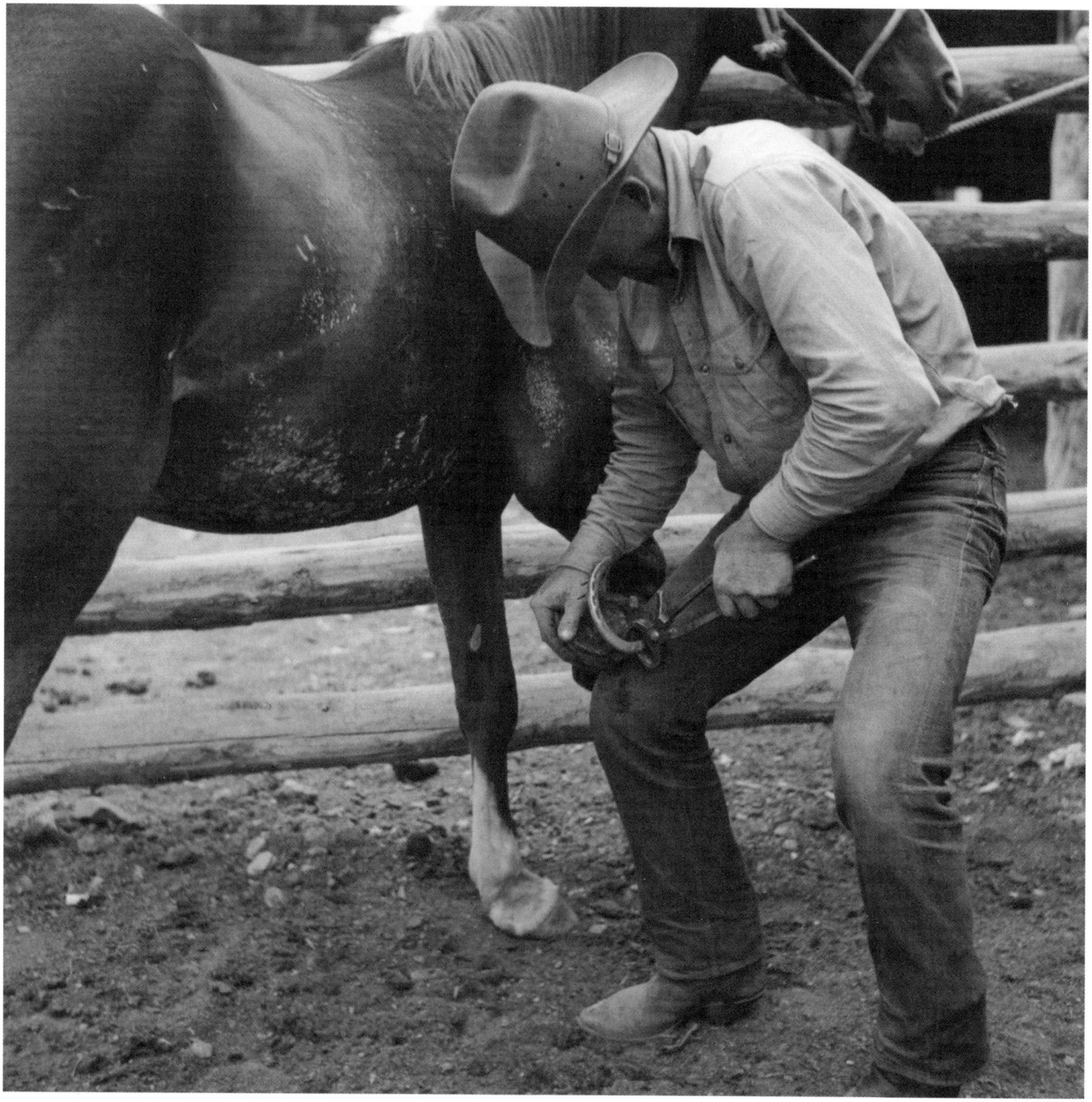

In the corral at Quarter Circle U, Brewster-Arnold Ranch Company. Removing horseshoes at the end of the summer season before turning the horses out on the range for the winter. Birney, Montana. Marion Post Wolcott, August 1941. *LC-USF34-058567-E DLC*

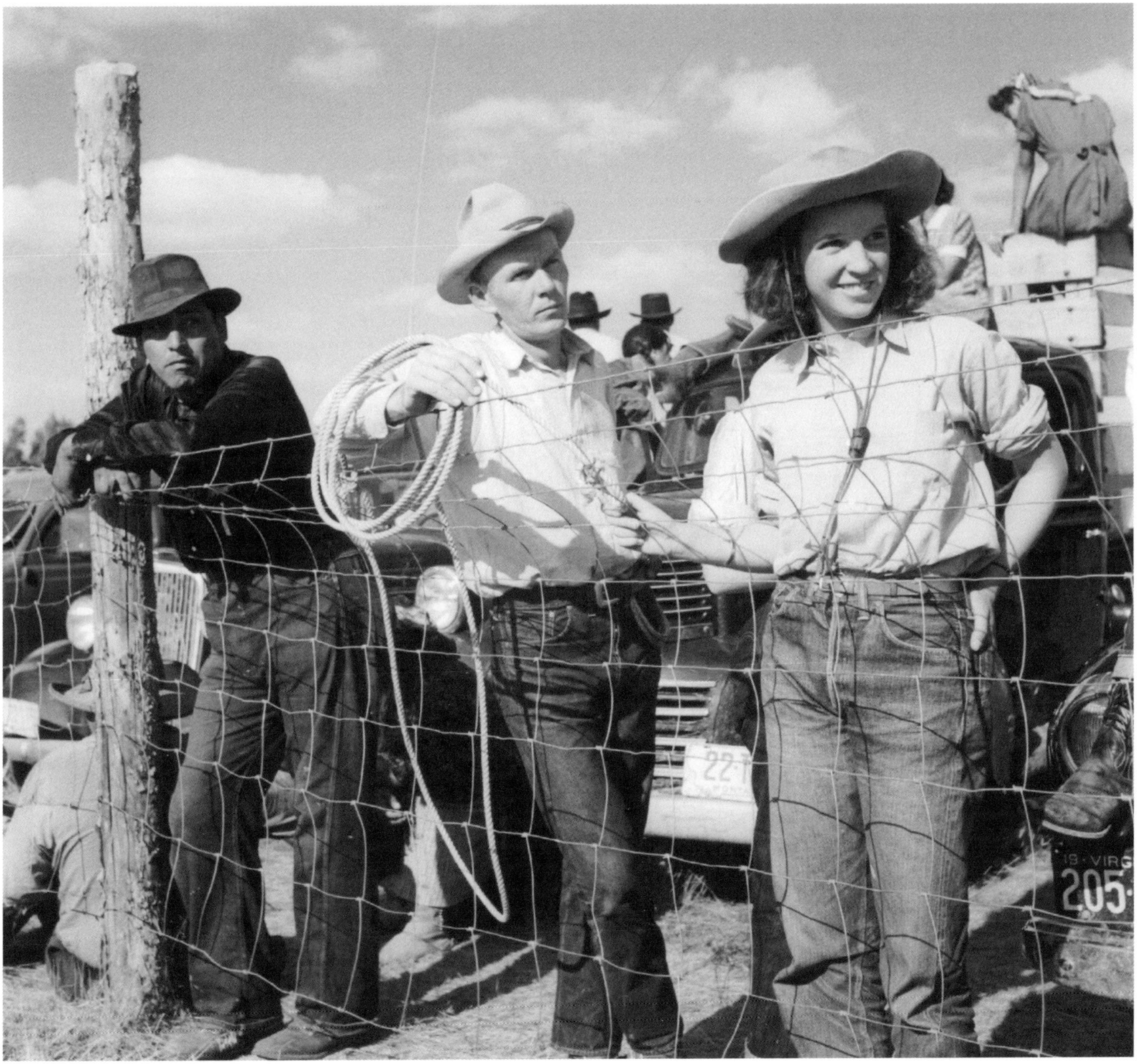

Dudes and cowboy from Quarter Circle U Ranch at Crow Indian Fair, Crow Agency, Montana. Marion Post Wolcott, September 1941.
LC-USF34-058892-E DLC

DUDE RANCHES

The first dude ranches were established following the Civil War and the cattle boom of the 1880s. Howard Eaton's Custer Trail Ranch, in the Dakota Badlands, was perhaps the earliest dude ranch. Eaton wrote folks back East to spread the word about the ranch, and soon Easterners who wanted to experience the vanishing frontier and get a taste of cowboy and ranch life were riding the new trains west. Both Theodore Roosevelt and Western-fiction writer Owen Wister were first introduced to the West at dude ranches. The cattle industry was struggling in the 1920s and ranchers saw the dude business as an additional source of income. By the 1920s, there were more than sixty dude ranches in Wyoming and Montana alone and numerous others elsewhere. The National Dude Ranchers' Association was created in 1926 and promoted the ranches in the East. As the Depression hit, ranchers needed the extra income even more. In 1934, the University of Wyoming offered its first Bachelor of Science degree in "recreational ranching." By 1936 as many as 356 ranches were in operation.

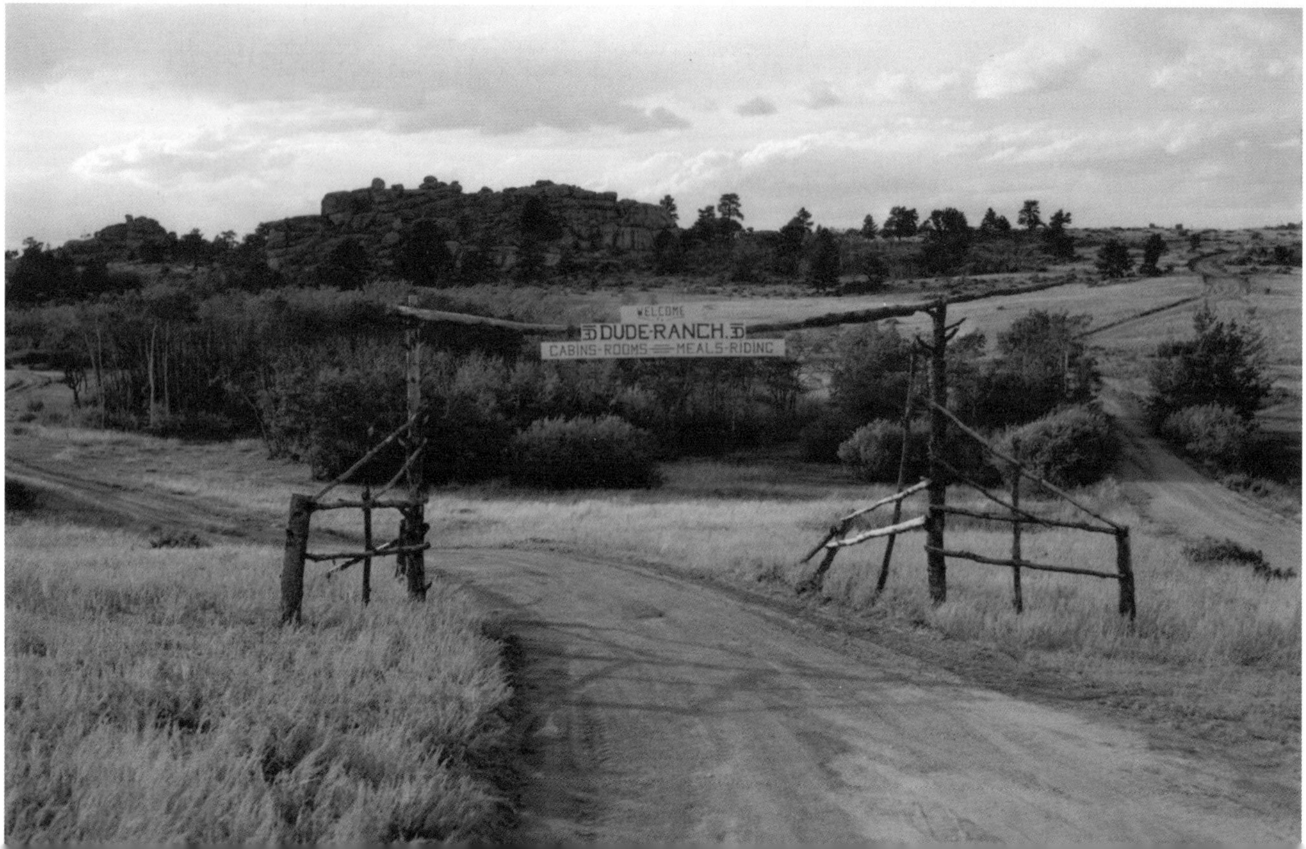

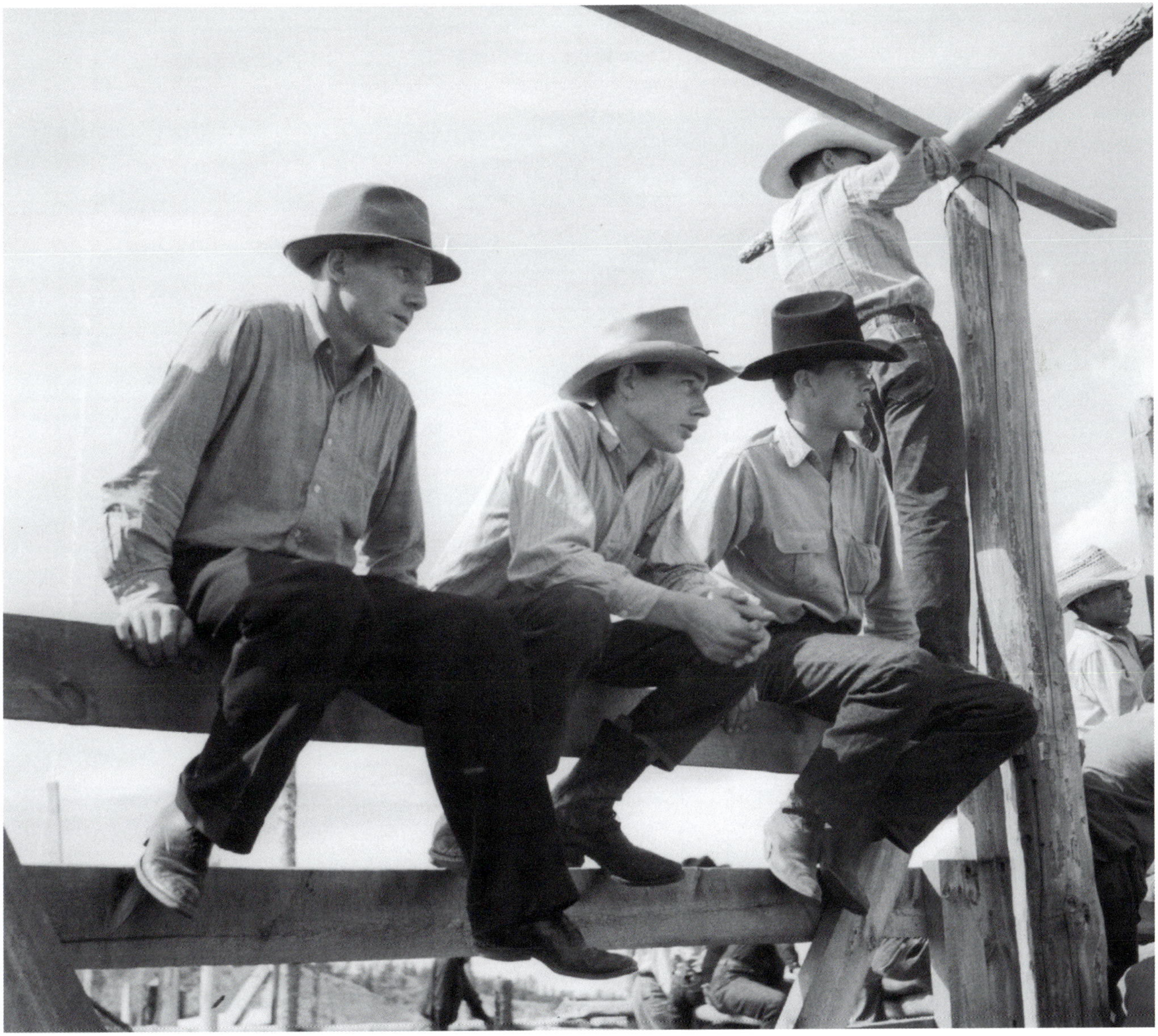

OPPOSITE: Gate entrance to Bar ED dude ranch. Buford, Wyoming. Marion Post Wolcott, September 1941.
LC-USF34-059367-D DLC

ABOVE: Boys from nearby ranches at the rodeo. Ashland, Montana. Marion Post Wolcott, September 1941.
LC-USF34-058897-E DLC

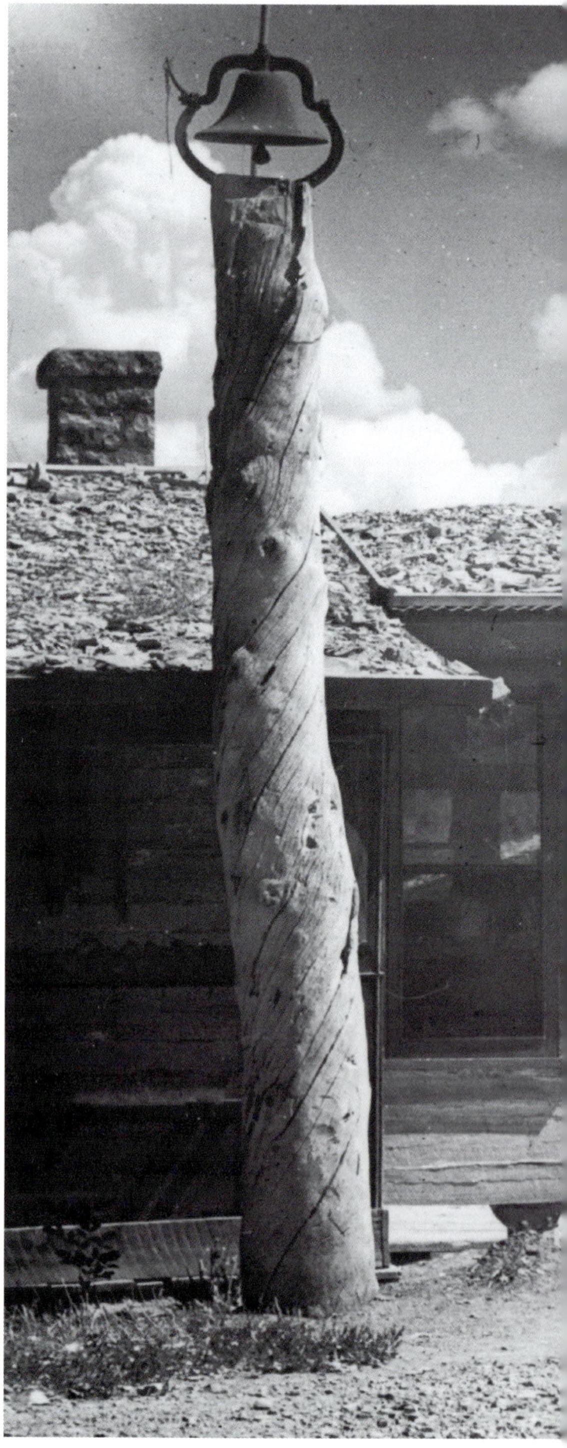

Old Ranch house now cowboy's quarters at Quarter Circle U, Brewster-Arnold Ranch Company. Birney, Montana. Marion Post Wolcott, August 1941.
LC-USF347-058497-D DLC

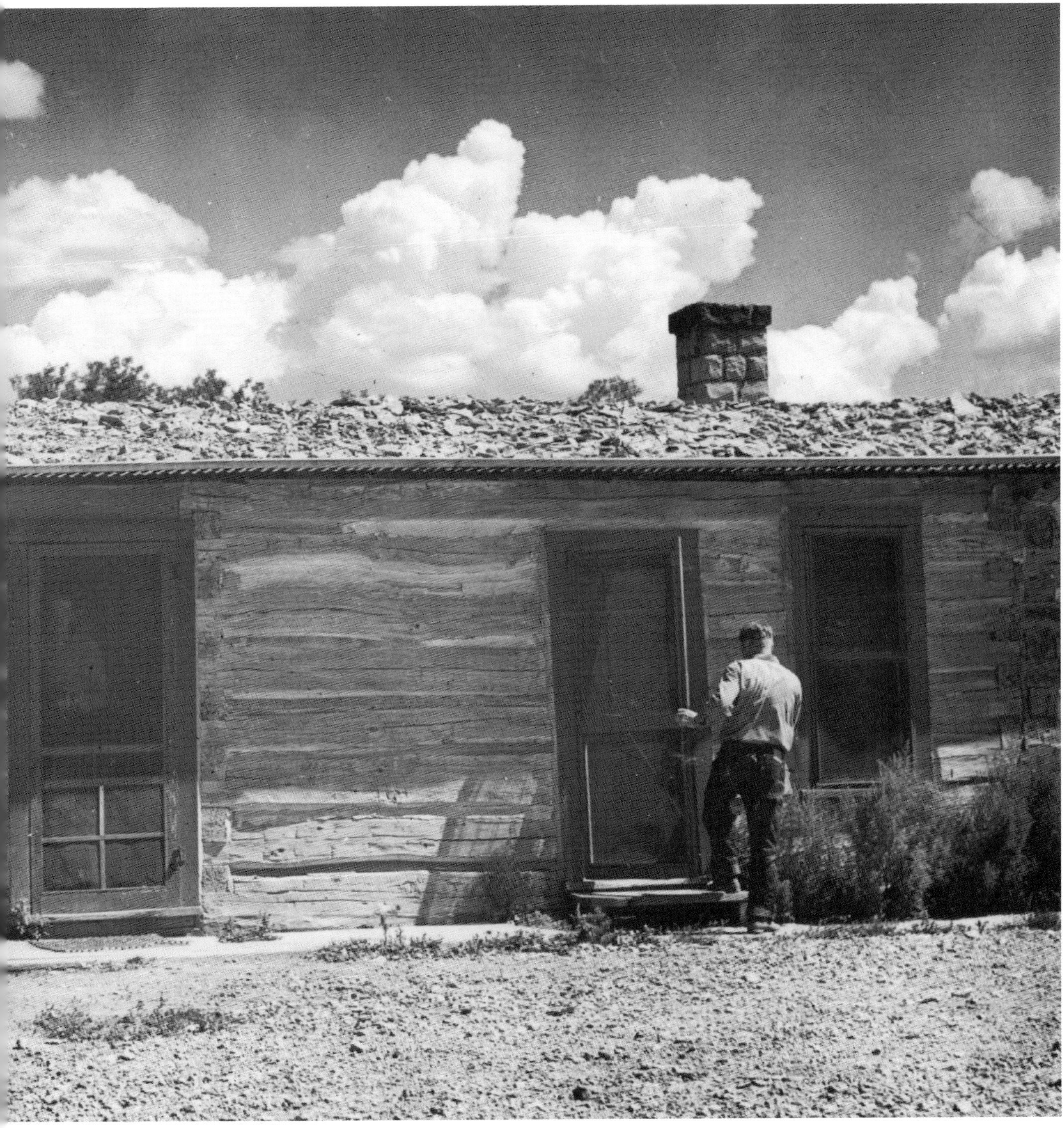

RODEO

Rodeos began as informal competitions among working cowboys to see who could better perform tasks that were part of their everyday work, such as bronc riding and roping. Often the cowboys on one ranch would compete against the cowboys of a neighboring ranch. At the beginning of the twentieth century, some of these informal competitions began to be held in front of paying audiences and their popularity spread. Many towns around the West began to sponsor local rodeos. Cowboys would pay money into a prize pool to compete.

By the time FSA photographers were documenting rodeos in the 1930s and 1940s, informal rodeo competitions had begun to develop into an organized sport. The first rodeo organization, the Rodeo Association of America, was formed in 1929. It formulated uniform rules for the burgeoning sport that allowed for the determination of winners and champions. Because of dissatisfaction with unfair practices in judging and payment, rodeo cowboys formed their own organization, the Cowboys' Turtle Association, in 1936, the same year the FSA photography project was initiated. The name was changed in 1945 to the Rodeo Cowboy's Association and in 1975 to the Professional Rodeo Cowboy's Association. Today more than 600 rodeos are held around the country each year.

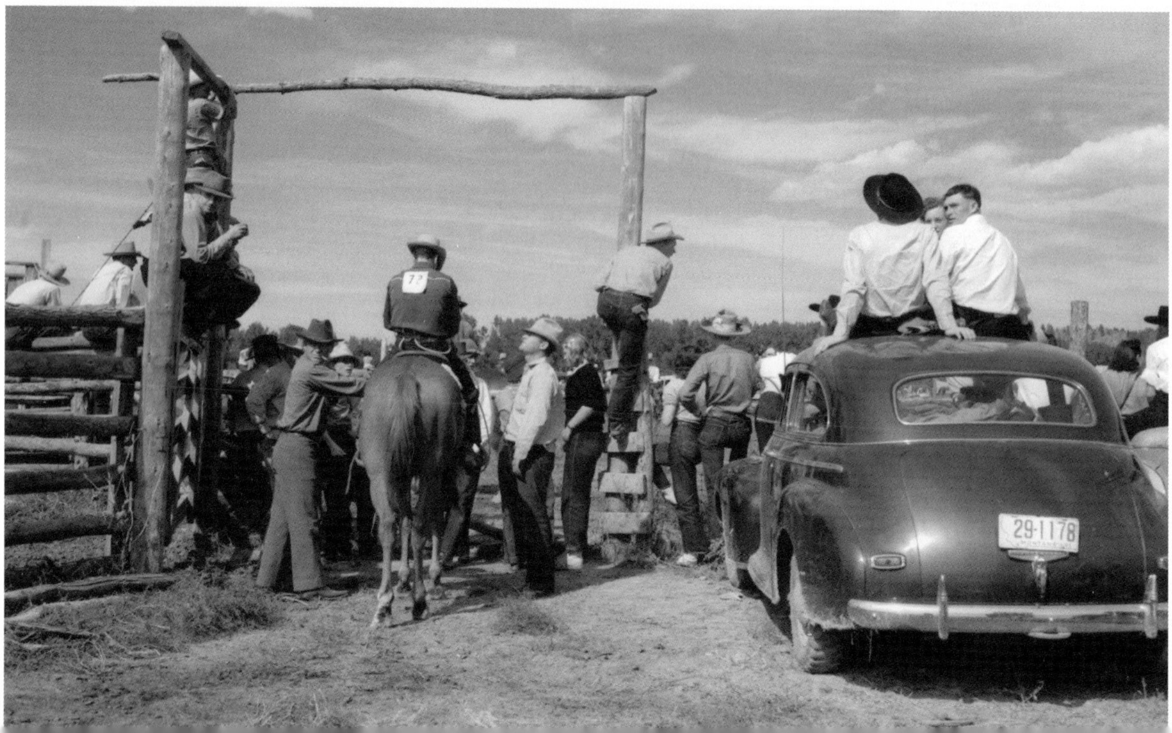

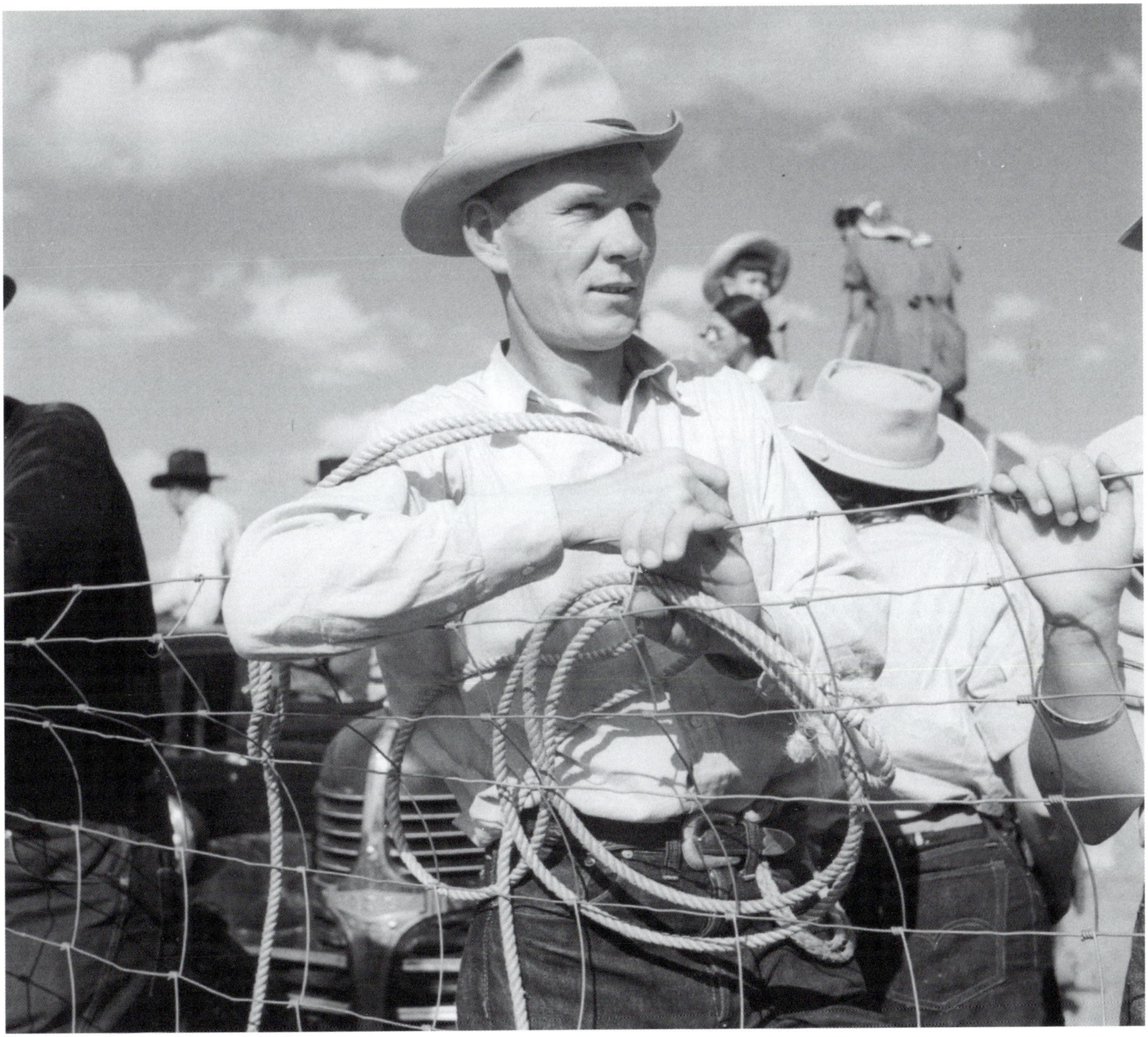

OPPOSITE: Cowboys and spectators at the annual rodeo. Ashland, Montana. Marion Post Wolcott, August 1941.
LC-USF34-058684-D DLC

ABOVE: Cowboy from Quarter Circle U Ranch at Crow Indian Fair. Crow Agency, Montana. Marion Post Wolcott, August 1941.
LC-USF34-058893-E DLC

JOHN COLLIER JR.

1913–1992

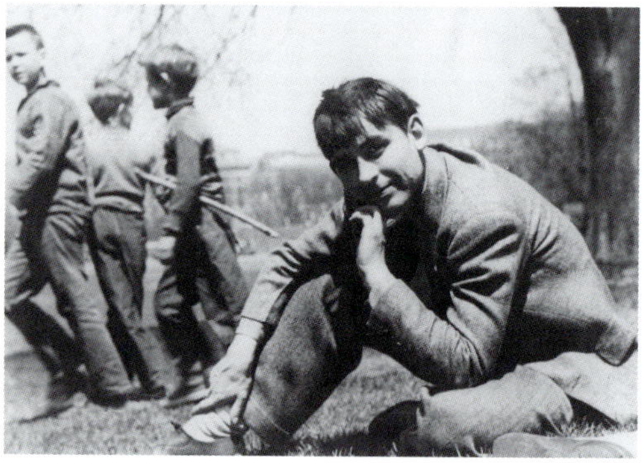

John Collier Jr., Farm Security Administration photographer, full-length portrait, seated on lawn with boys in the background. c. 1943.
LC-USZ62-120967 DLC

John Collier Jr., born May 22, 1913, in Sparkill, New York, was the son of sociologist John Collier, who was the head of the Bureau of Indian Affairs during the New Deal from 1933 to 1945. Because of his father's work, the family maintained ties to the area around the Taos Indian Pueblo in New Mexico, and John mostly grew up there and in the San Francisco Bay area. When he was eight years old, he was seriously injured in an automobile accident that resulted in brain damage and associated learning disabilities and hearing loss that prevented him from successfully completing schooling beyond a sixth-grade level.

When Collier was twelve, he was apprenticed to Maynard Dixon, a well-known painter, to afford him opportunities outside of formal schooling. Dixon's wife, photographer Dorothea Lange, introduced him to photography. In the early 1930s, he served as an informal guide to the photographer Paul Strand while Strand was in the Taos region, but he continued to attempt a career in painting and writing through the mid-1930s. After a brief enrollment at the California School of Fine Arts (now the San Francisco Art Institute) where he took more painting classes, he turned to photography. He was largely self-trained, except for some instruction in studio techniques from Sara Higgins Mack. In 1939, after working for a period in San Francisco, he opened a photography studio in Taos, using what had been Paul Strand's darkroom and studio. In 1941, probably through the influence of Dorothea Lange, some of his photography from New Mexico came to the attention of Roy Stryker, who hired him to work for the Farm Security Administration. He stayed with the FSA until 1943 and documented many areas around the eastern United States and in northern New Mexico. In 1943, he married Mary Elizabeth Trumbull, who became a photographer herself and an important partner in his field projects. In 1967, he published *Visual Anthropology: Photography as a Research Method*, which has remained a classic in the field. In the fields of visual anthropology and visual sociology, Collier is recognized as a major methodological pioneer, in particular for the development of "photo-elicitation" techniques in which photographs are used systematically in interviews to elicit information and insight. From 1961 to 1983, he taught in the Anthropology Department at San Francisco State University, where he became a full professor.

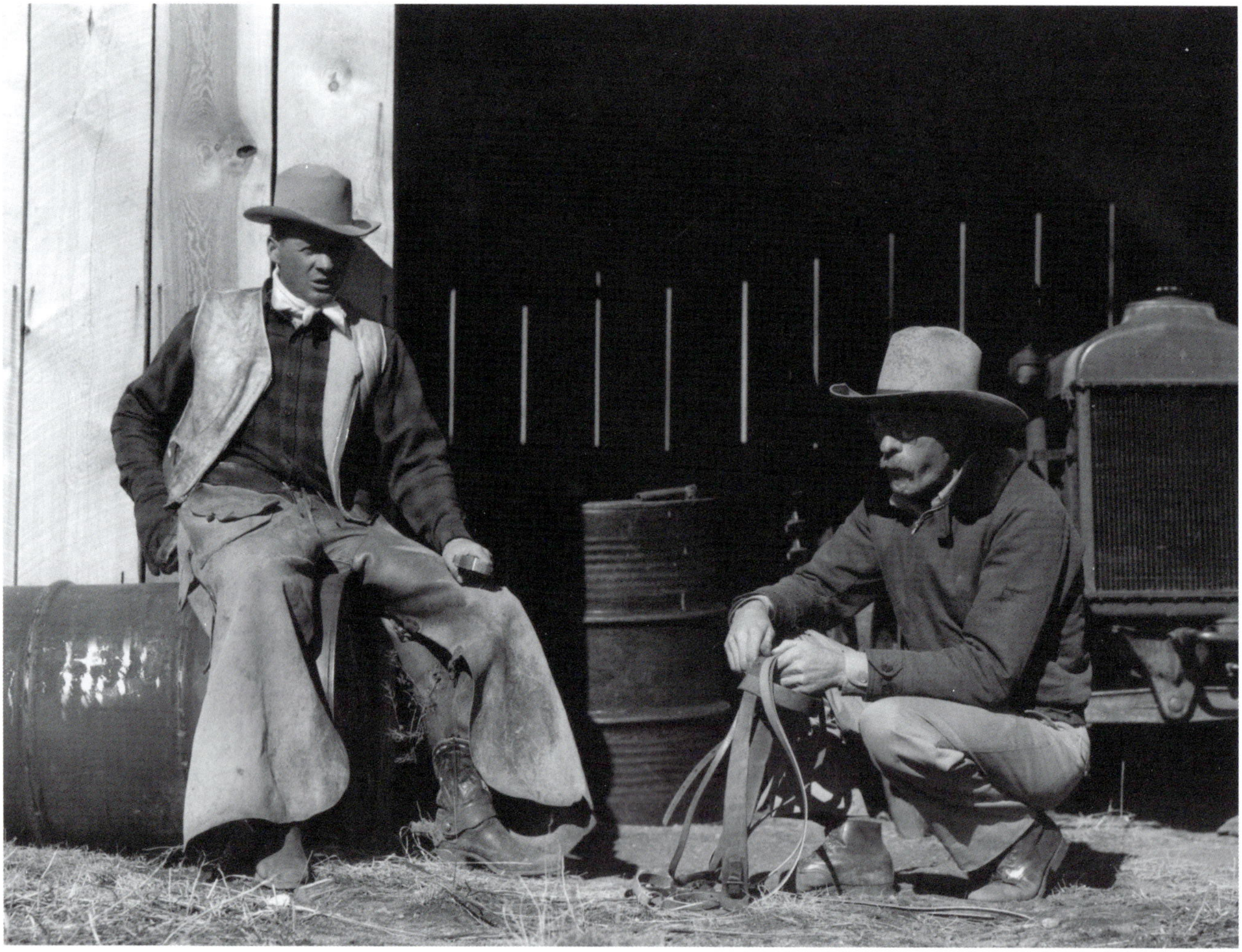

ABOVE: John Mutz and George Turner, ranchers, talking things over. Moreno Valley, Colfax County, New Mexico. John Collier, February 1943.
LC-USW3-018676-C DLC

OPPOSITE: A Spanish-American rancher. Chacon, Mora County, New Mexico. John Collier, January 1943.
LC-USW3-015281-C DLC

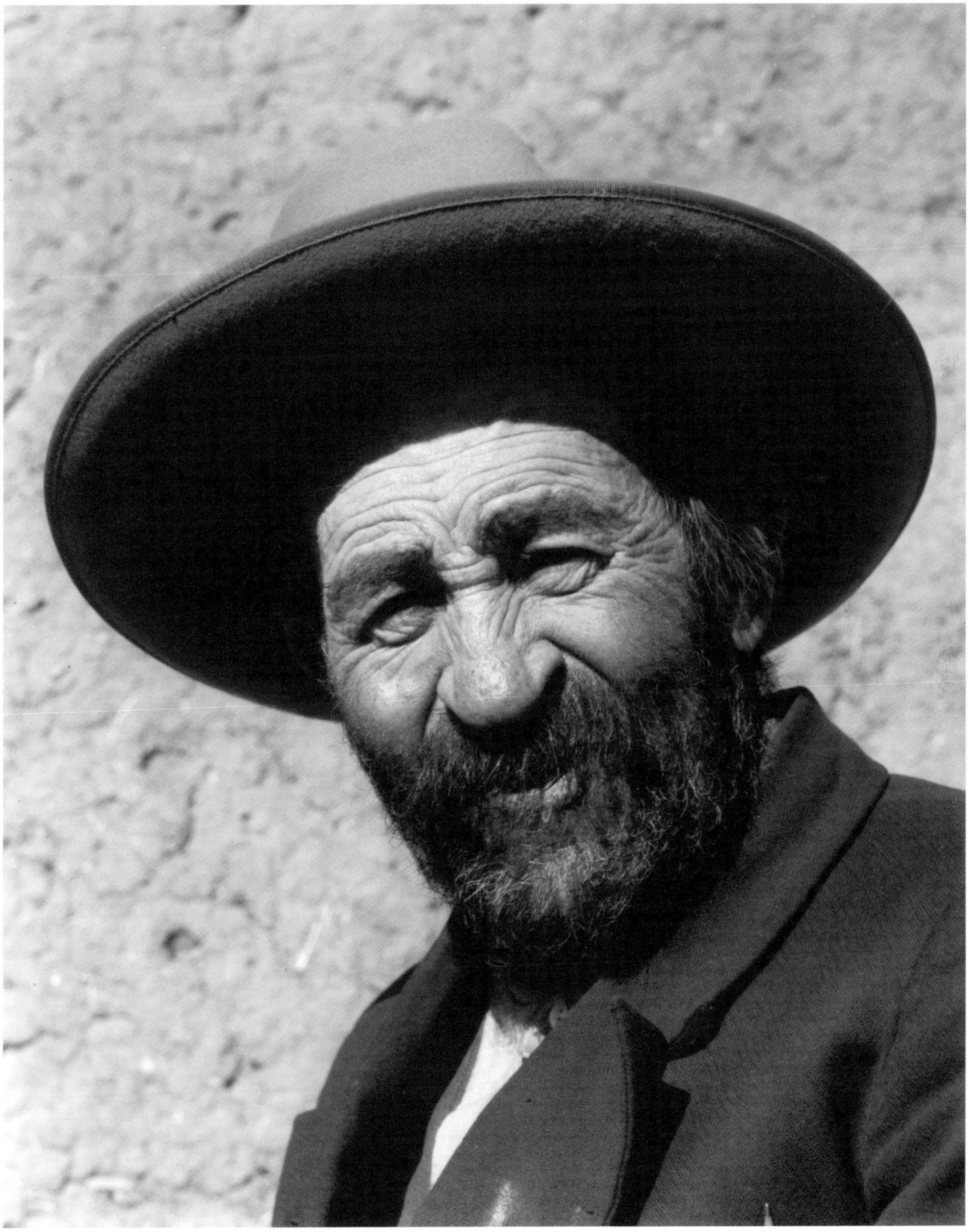

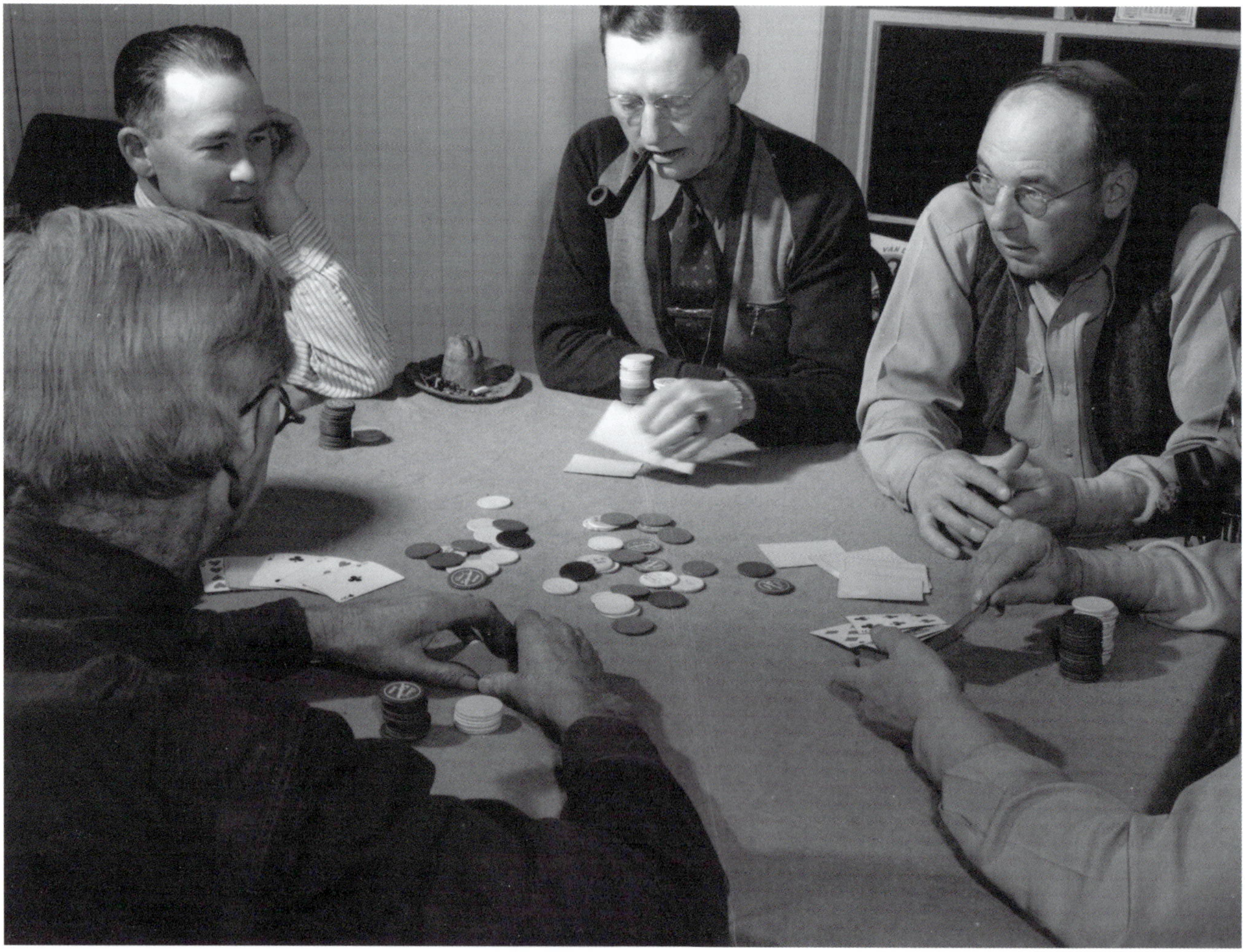

ABOVE: A poker party at George Turner's ranch. Moreno Valley, Colfax County, New Mexico. John Collier, January 1943. *LC-USW3-018756 DLC*

OPPOSITE: A Spanish-American rancher. Chacon, Mora County, New Mexico. John Collier, January 1943. *LC-USW3-015318-C DLC*

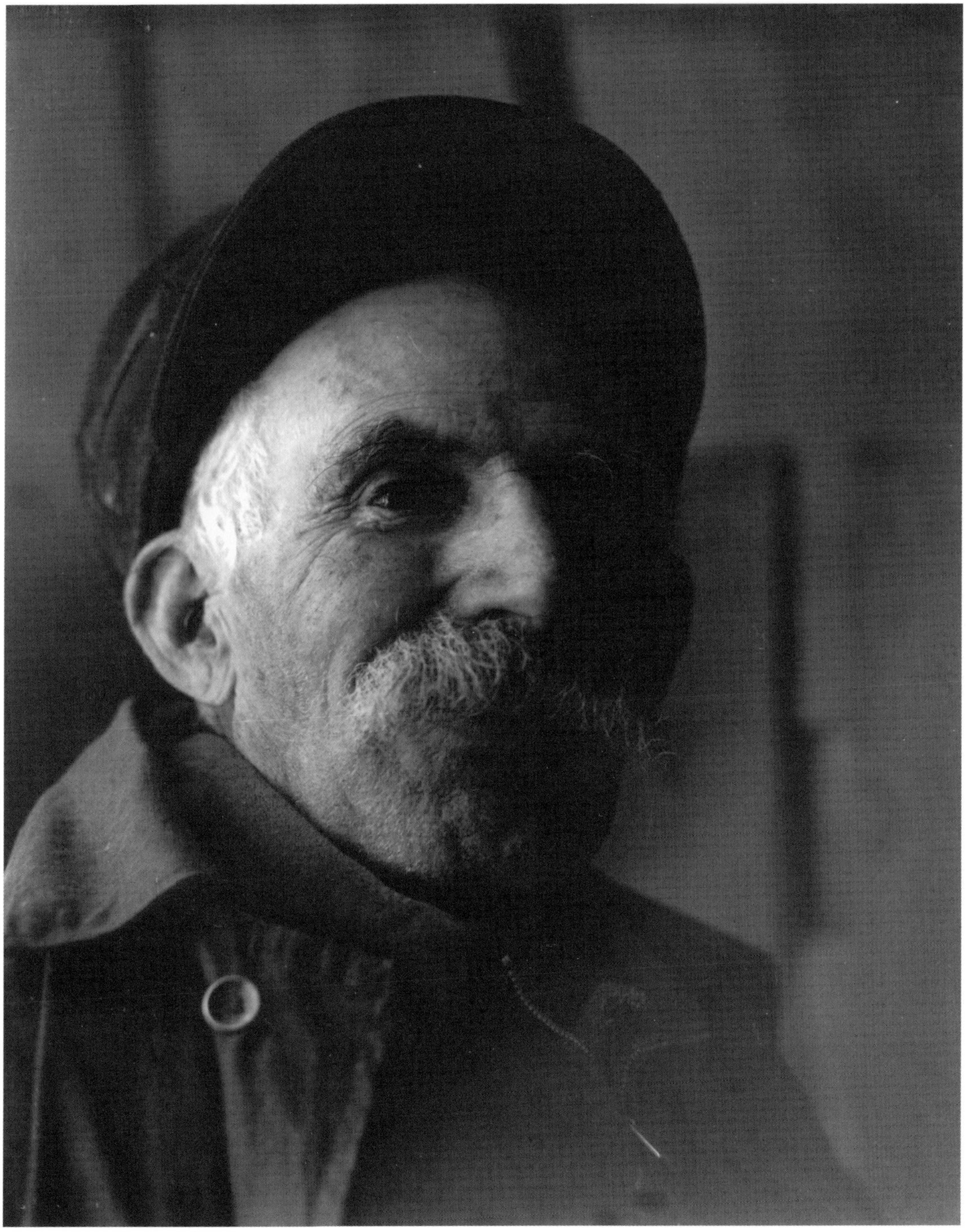

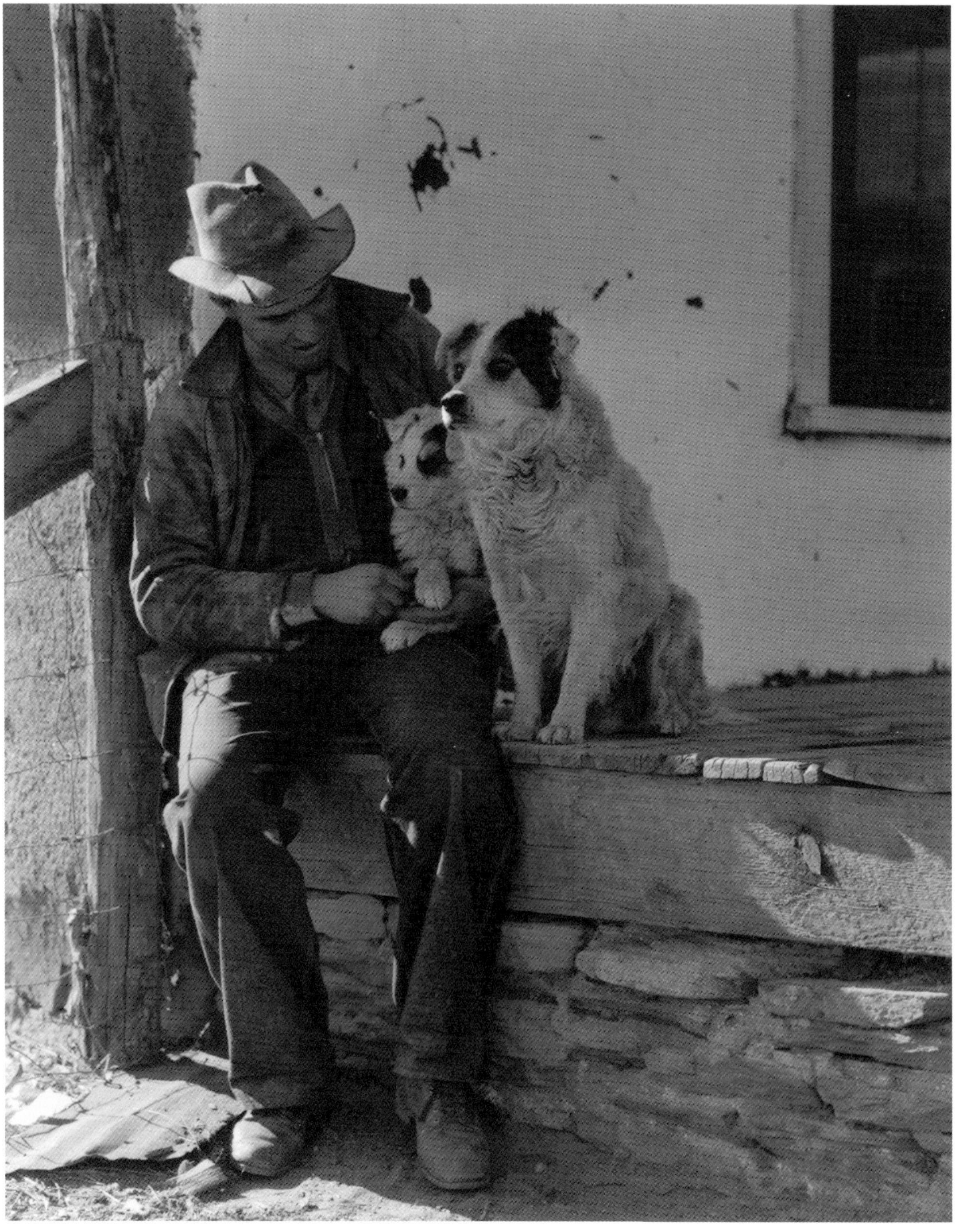

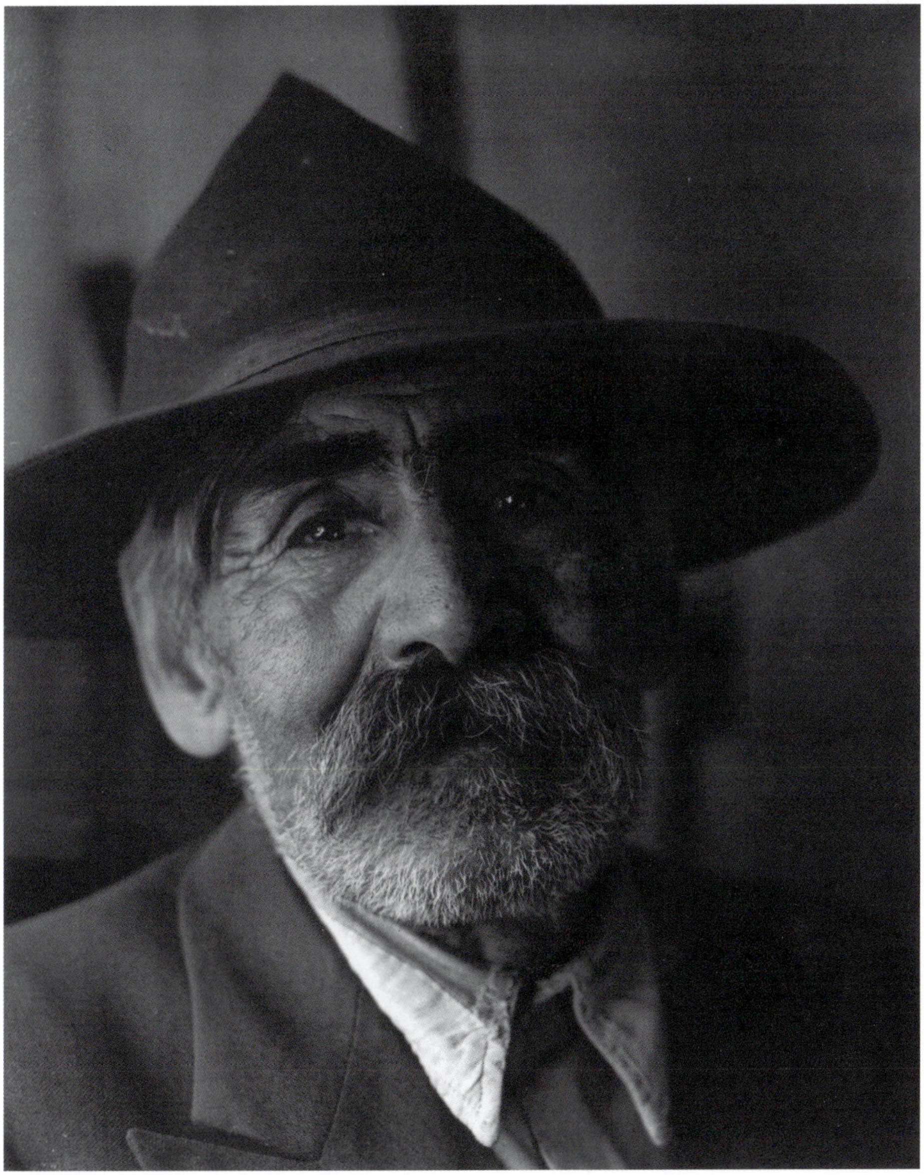

OPPOSITE: An Anglo rancher. Near Mora, New Mexico. John Collier, January 1943.
LC-USW3-017932-C DLC

ABOVE: A Spanish-American rancher. Chacon, Mora County, New Mexico. John Collier, January 1943.
LC-USW3-015317-C DLC

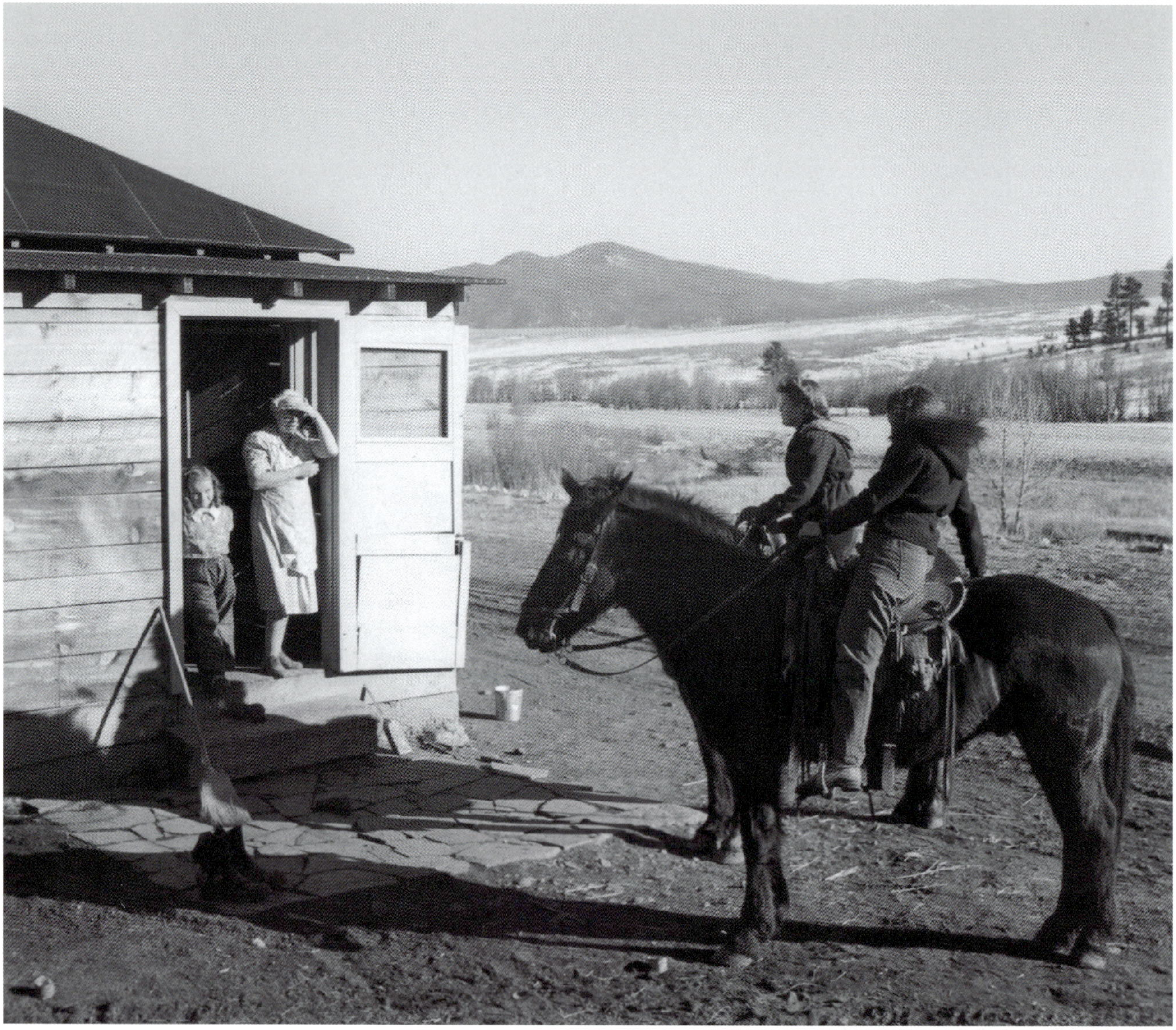

ABOVE: George Mutz's daughters calling on a nearby ranch. Moreno Valley, Colfax County, New Mexico. John Collier, February 1943.
LC-USW3-018848-E

OPPOSITE: William Heck, a rancher, at his home. Moreno Valley, Colfax County, New Mexico. John Collier, February 1943.
LC-USW3-018953-E

RADIO

One of the communications innovations that had a major impact on ranch life was the radio. Radio service began to grow in the early 1920s, and initially many stations went on the air specifically to serve isolated ranchers and farmers in rural areas, bringing them weather forecasts, morning and noon market reports, and information on new agricultural and animal husbandry practices. Many farmers had battery-powered radios before they had electricity. In 1921 there was only one radio for every 500 households in the United States, but by 1926 there was one radio for every six households. Radio connected ranches with the popular culture of the times, bringing rural and urban people closer together as they listened to the same national news, music, and radio shows.

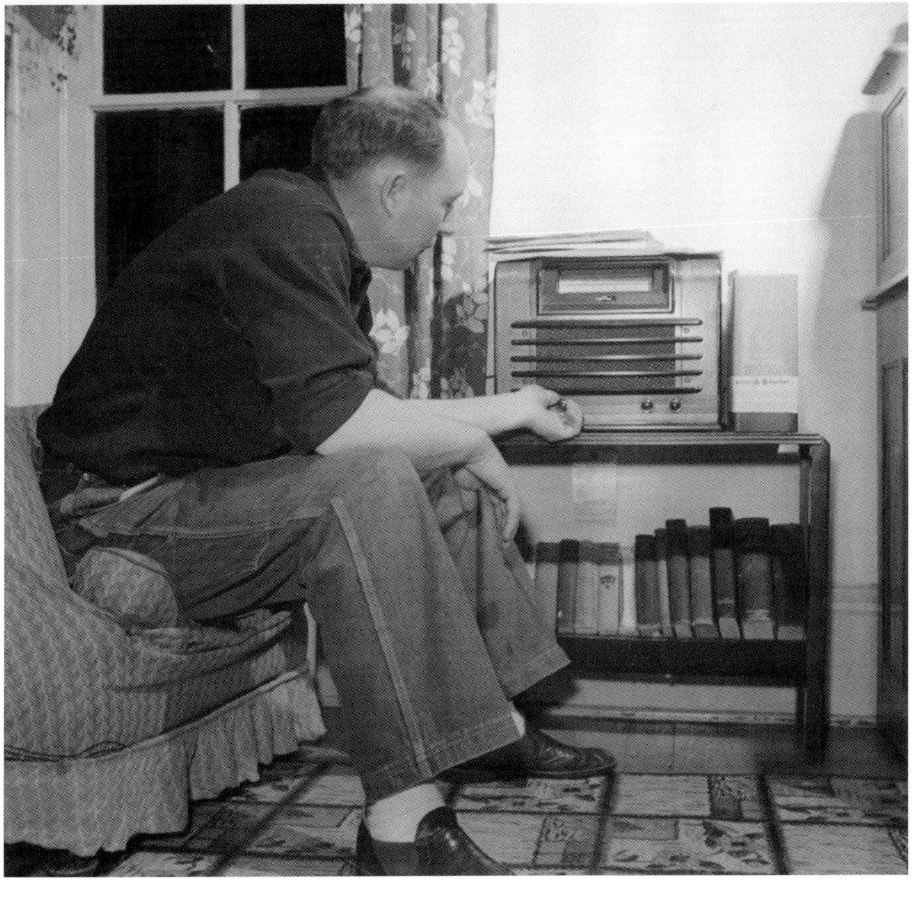

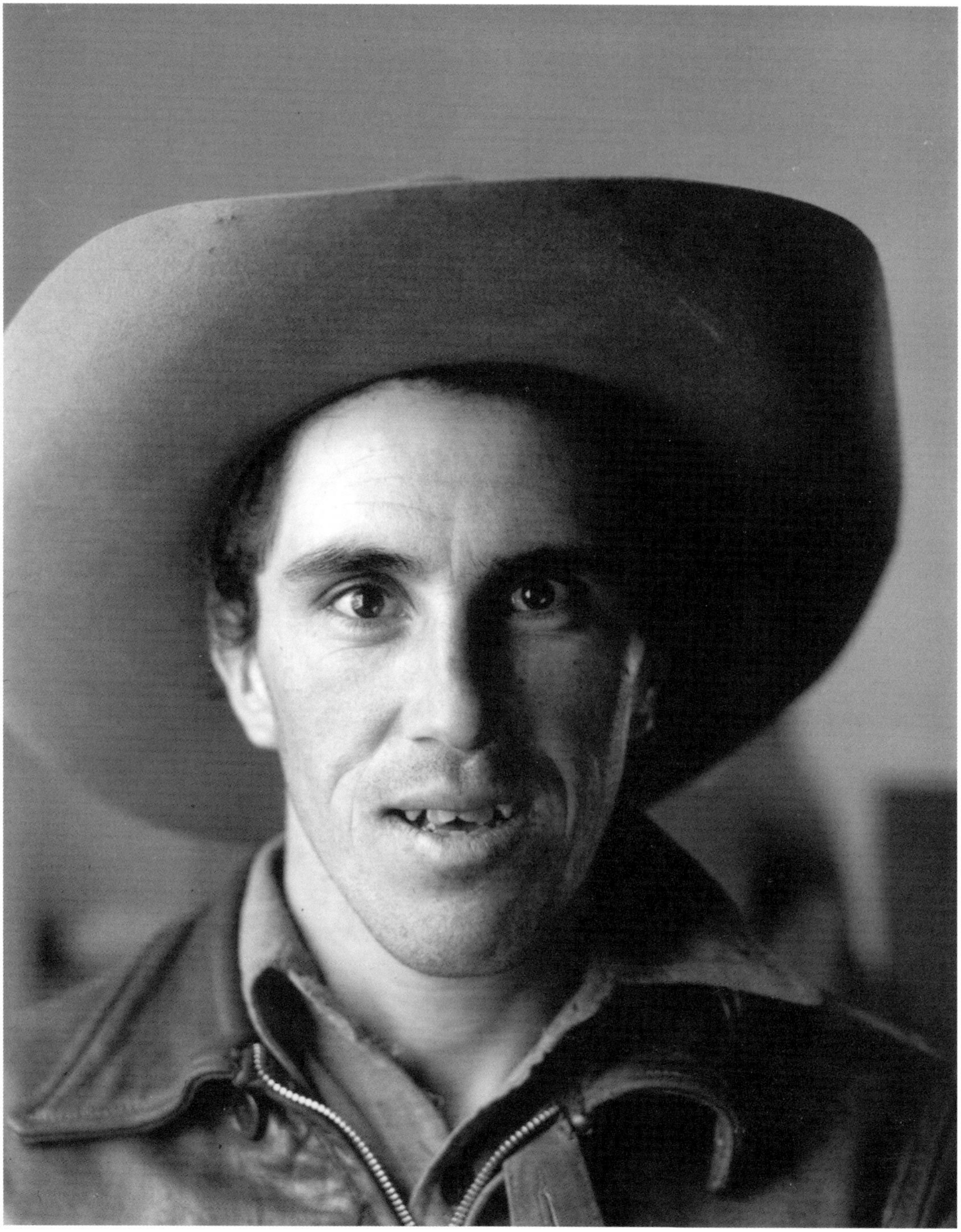

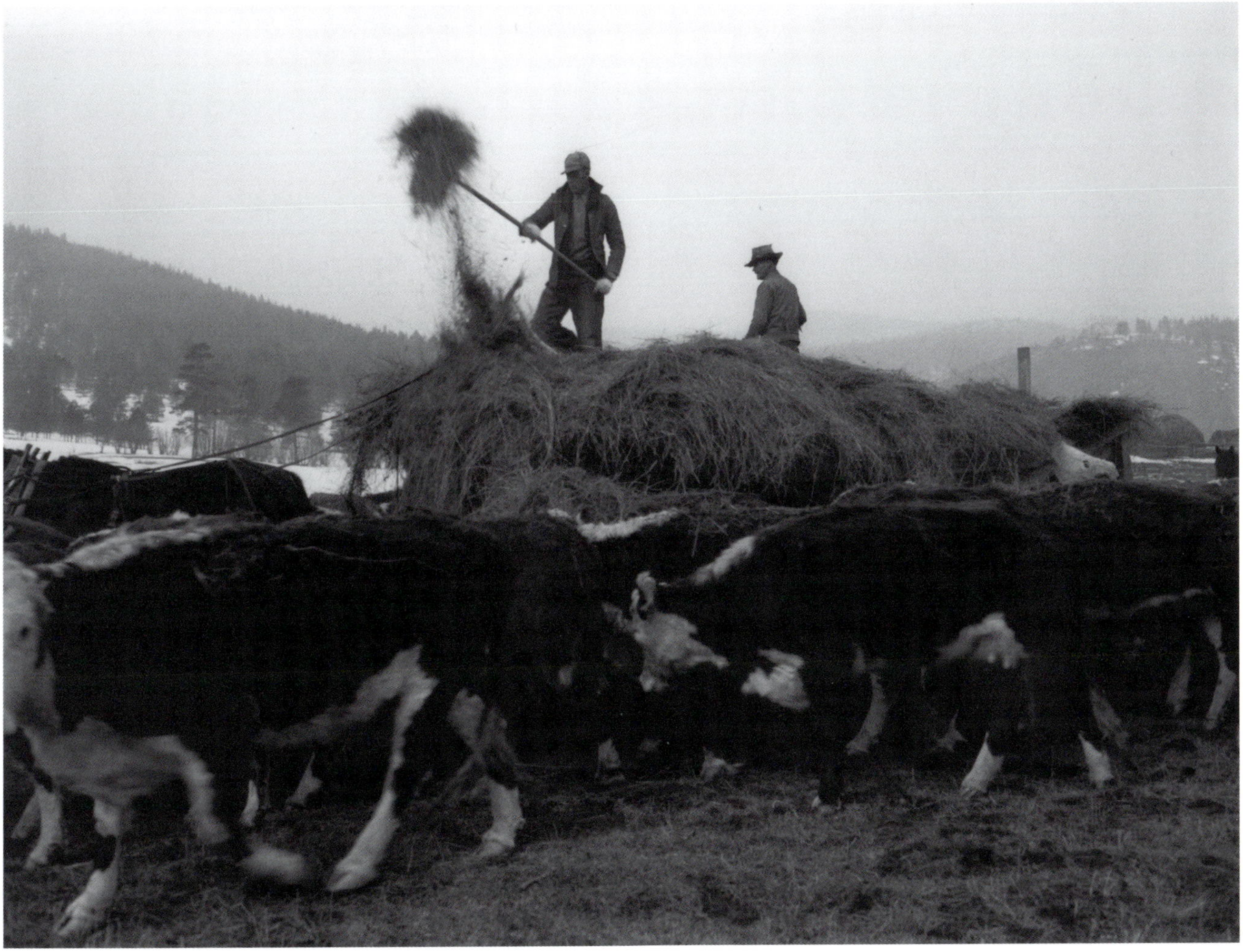

OPPOSITE: A Spanish-American rancher. Chacon, Mora County, New Mexico. John Collier, January 1943.
LC-USW3-015279-C DLC

ABOVE: Cattle winter feeding on George Mutz's ranch. Moreno Valley, Colfax County, New Mexico. John Collier, January 1943.
LC-USW3-018700-C

DOROTHEA LANGE 1895–1965

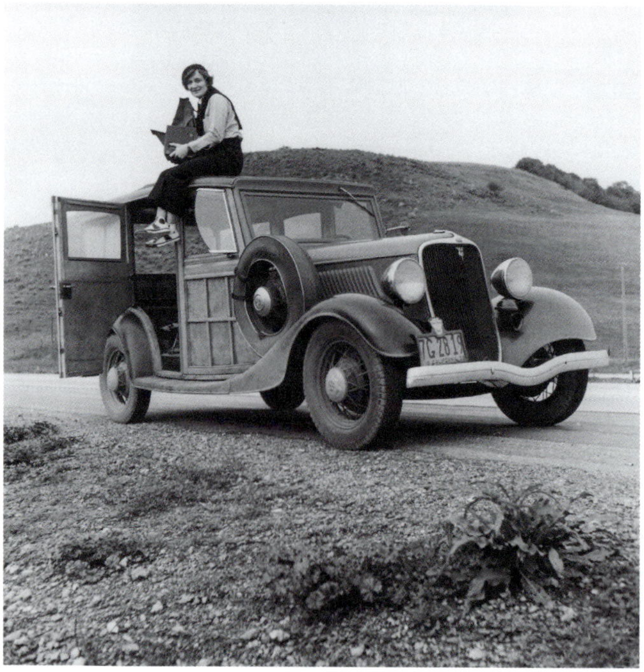

Dorothea Lange in California. February 1936.
LC-USF34-002392-E

Born Dorothea Nutzhorn in Hoboken, New Jersey, on May 26, 1895, she was the daughter of Joan Lange and Henry Nutzhorn. Dorothea developed polio in 1902, at age seven, which left her with a permanent limp. When she was twelve years old, her father abandoned her and her mother, leading her to drop her middle and last names and adopt her mother's maiden name. She was educated in photography in New York City, in a class taught by Clarence H. White. She was informally apprenticed to several New York photography studios, including that of Arnold Genthe. In 1918, she moved to San Francisco, and by the following year, she had opened a successful portrait studio. She lived across the bay in Berkeley for the rest of her life. In 1920, she married the painter Maynard Dixon. With the onset of the Great Depression, Lange turned her camera lens from the studio to the street. Her studies of unemployed and homeless people, starting with *White Angel Breadline* (1933), which depicted a lone man turned away from the crowd in front of a soup kitchen run by a widow known as the White Angel, captured the attention of local photographers. They brought her work to the attention of Roy Stryker who hired her as a photographer for the FSA, where she worked until 1939. Known most widely for powerful documentation of the Dust Bowl and Okie migrants, Lange did take a handful of photos of cowboys, and those are included here. She received the Guggenheim prize in 1941 for her iconic 1936 photo *Migrant Mother*, which did much to bring the plight of migrants to the attention of the country.

Stump rancher and wife. They and their seven children are making a new start on cut-over land. Priest River Peninsula, Bonner County, Idaho. Dorothea Lange, October 1939. *LC-USF34-021788-C DLC*

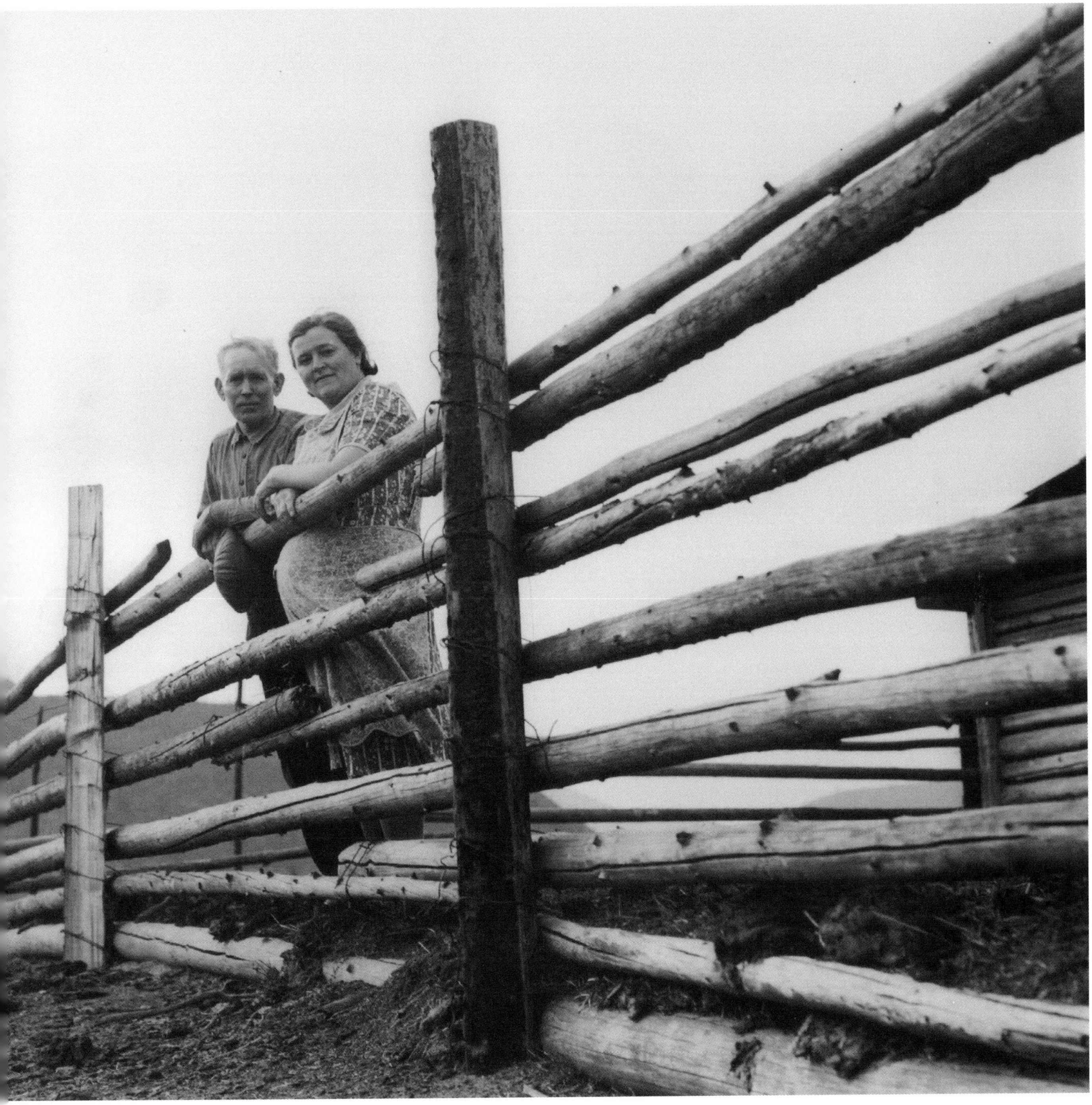

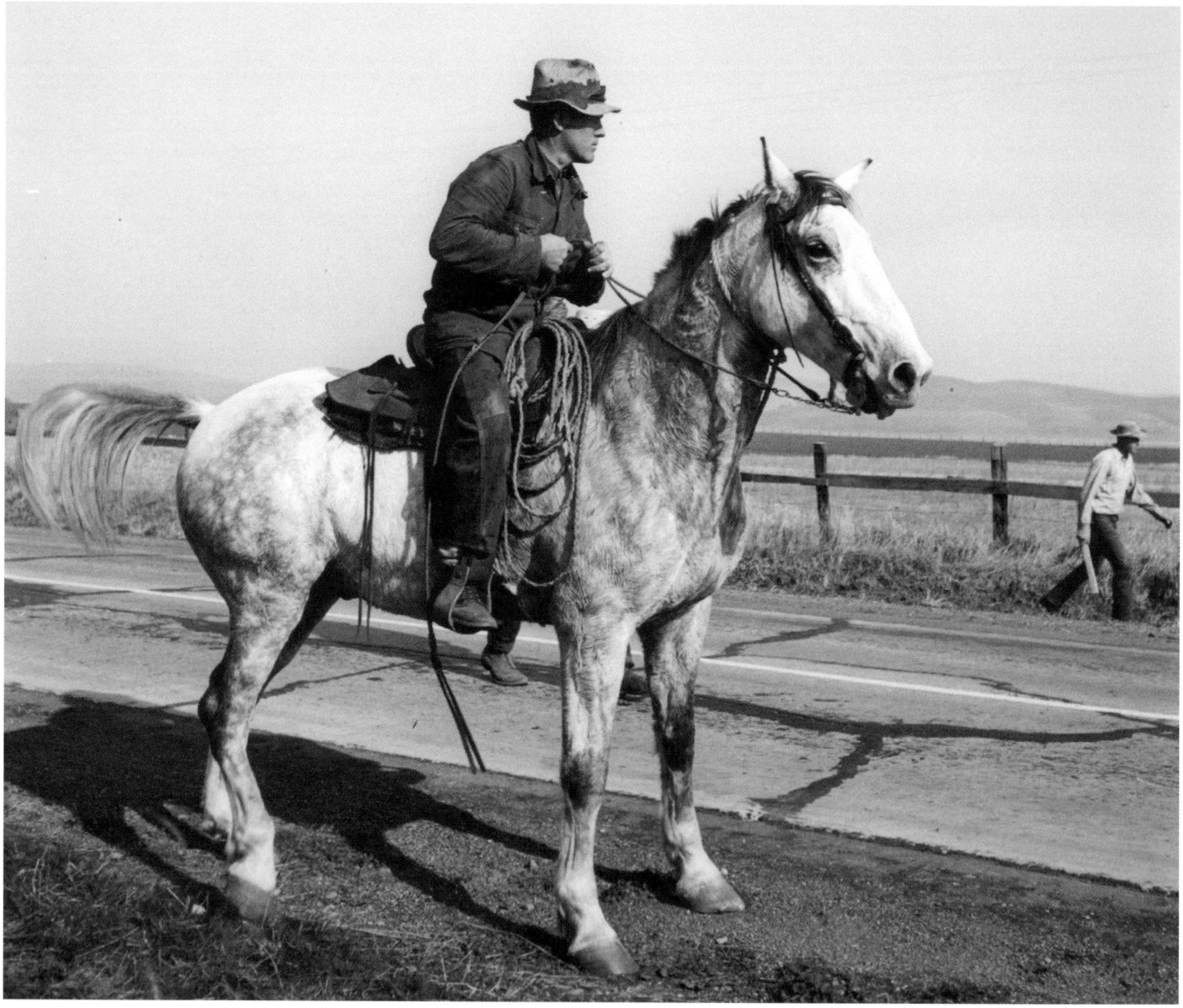

ABOVE: Cowboy bringing cattle in from the range. A common sight on California highways. Contra Costa County, California. Dorothea Lange, November 1938.
LC-USF34-018759-E DLC

OPPOSITE: Bringing cattle in from the range. A common sight on California highways. Contra Costa County, California. Dorothea Lange, November 1938.
LC-USF34-018756-E DLC

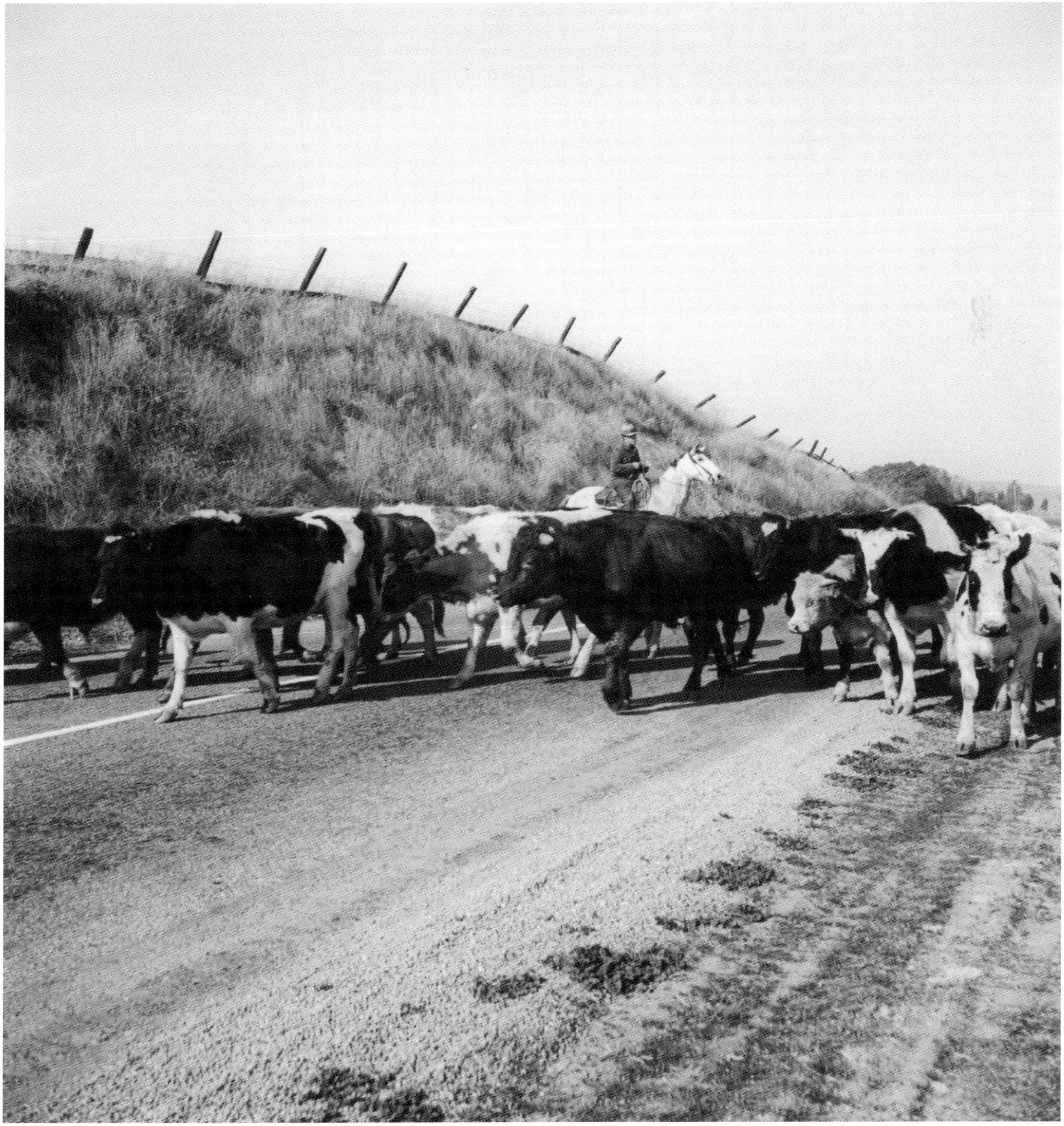

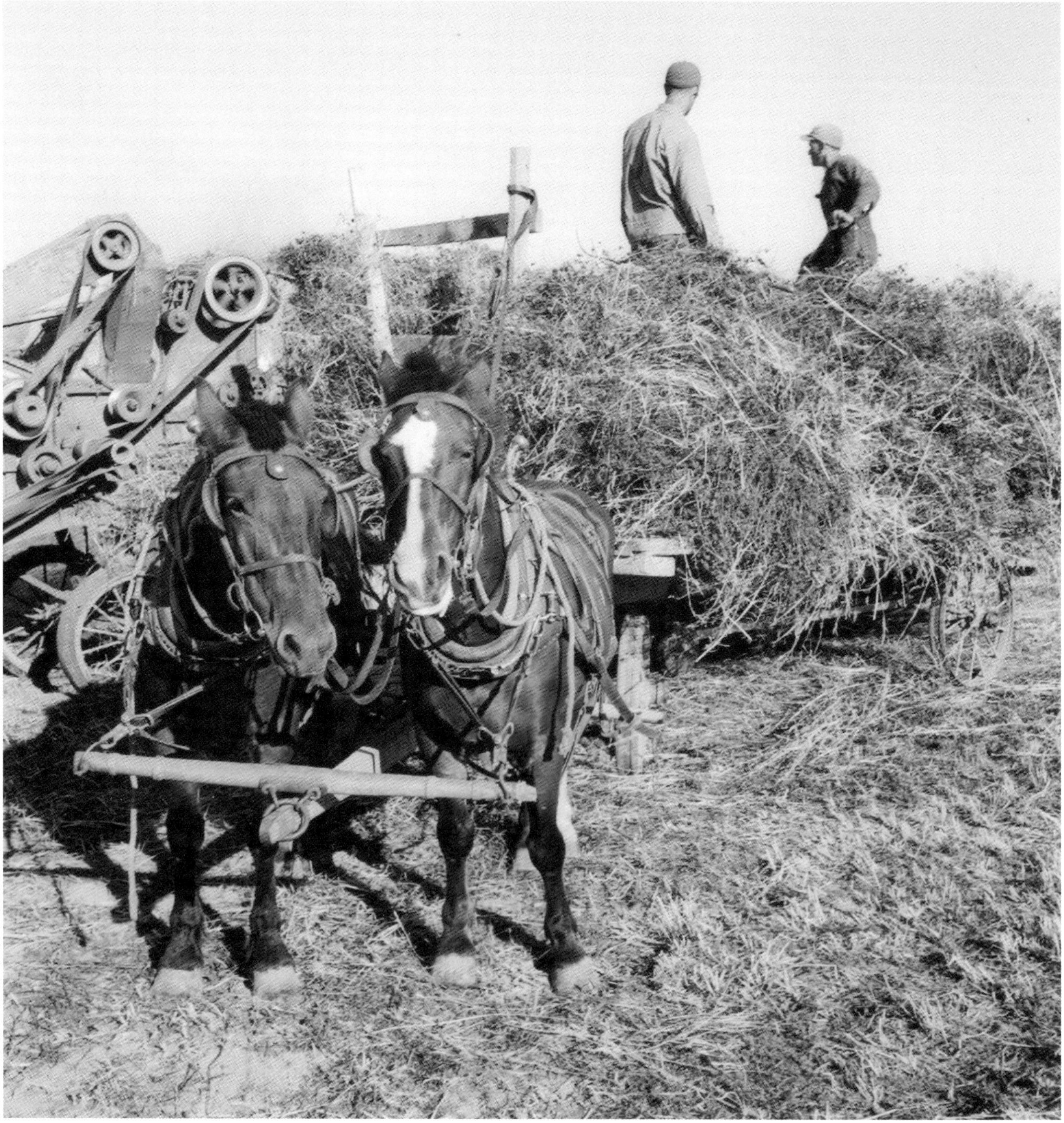

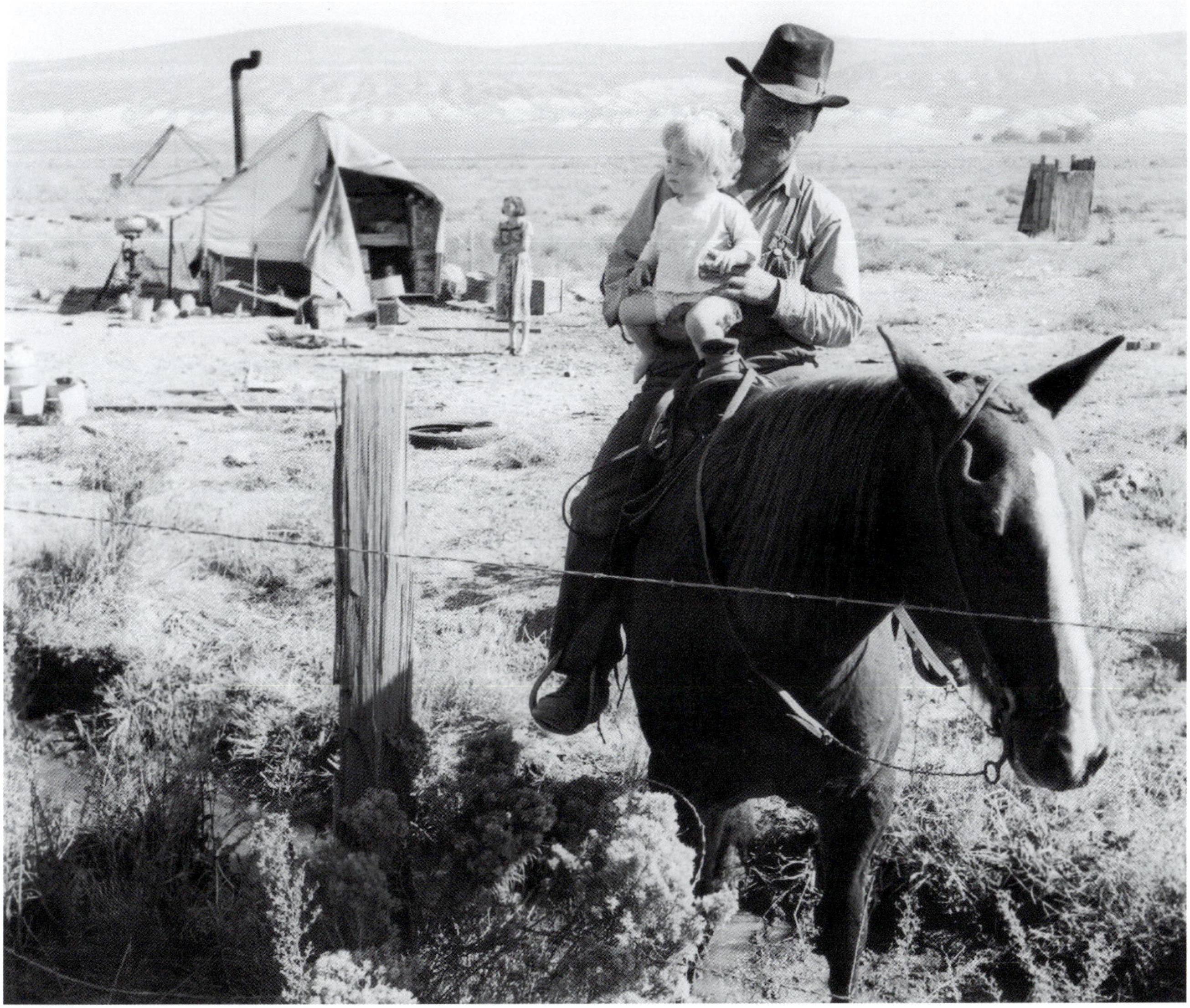

OPPOSITE: Threshing red clover for seed on old settler's ranch. Near Ontario, Malheur County, Oregon. Dorothea Lange, October 1939.
LC-USF34-021325-E DLC

ABOVE: The Fairbanks family has moved to three different places on the project in one year. Willow Creek area, Malheur County, Oregon. Dorothea Lange, October 1939.
LC-USF34-021263-E

JOHN VACHON 1914–1975

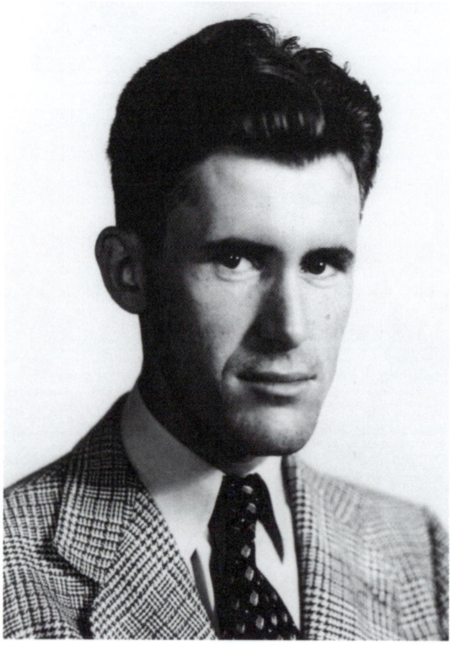

John F. Vachon, Farm Security Administration photographer, 1942.
LC-DIG-ppmsc-00194 DLC; LC-USF341-T01-014695-C DLC; LC-USF341-T01-014695-C

John Vachon came from his native Michigan in 1936 as a graduate student in English literature at Catholic University in Washington, D.C. Roy Stryker hired him at the Farm Security Administration and gave him the title of "assistant messenger," purportedly because he had no ambitions to become a photographer. By 1937, Vachon had developed an interest in photography, and Walker Evans, Arthur Rothstein, and Ben Shahn tutored him in photographic skills. He persuaded Stryker to let him document scenes in the District of Columbia that would make a contribution to the collection. After the Washington Post published some of his photographs, Stryker began to give him assignments. In October and November 1938, Vachon traveled to Nebraska on his first extensive solo trip. He photographed agricultural programs on behalf of the FSA's regional office and pursued an extra assignment from Stryker: the city of Omaha. Stryker continued sending Vachon on longer photo trips, although he did not acquire the title of junior photographer until 1940. In 1942, he was sent to Montana and the Dakotas for about four months. Vachon continued working for the FSA after it became the Office of War Information (OWI) in 1942. In 1947, Vachon moved to New York, where he became a member of the Photo League, contributing numerous book reviews to the newsletter Photo Notes and participating in the 1948 exhibition *This Is the Photo League.* After working for many years as a staff photographer at *Look* magazine, Vachon became a visiting professor at the Minneapolis College of Art and Design in 1974.

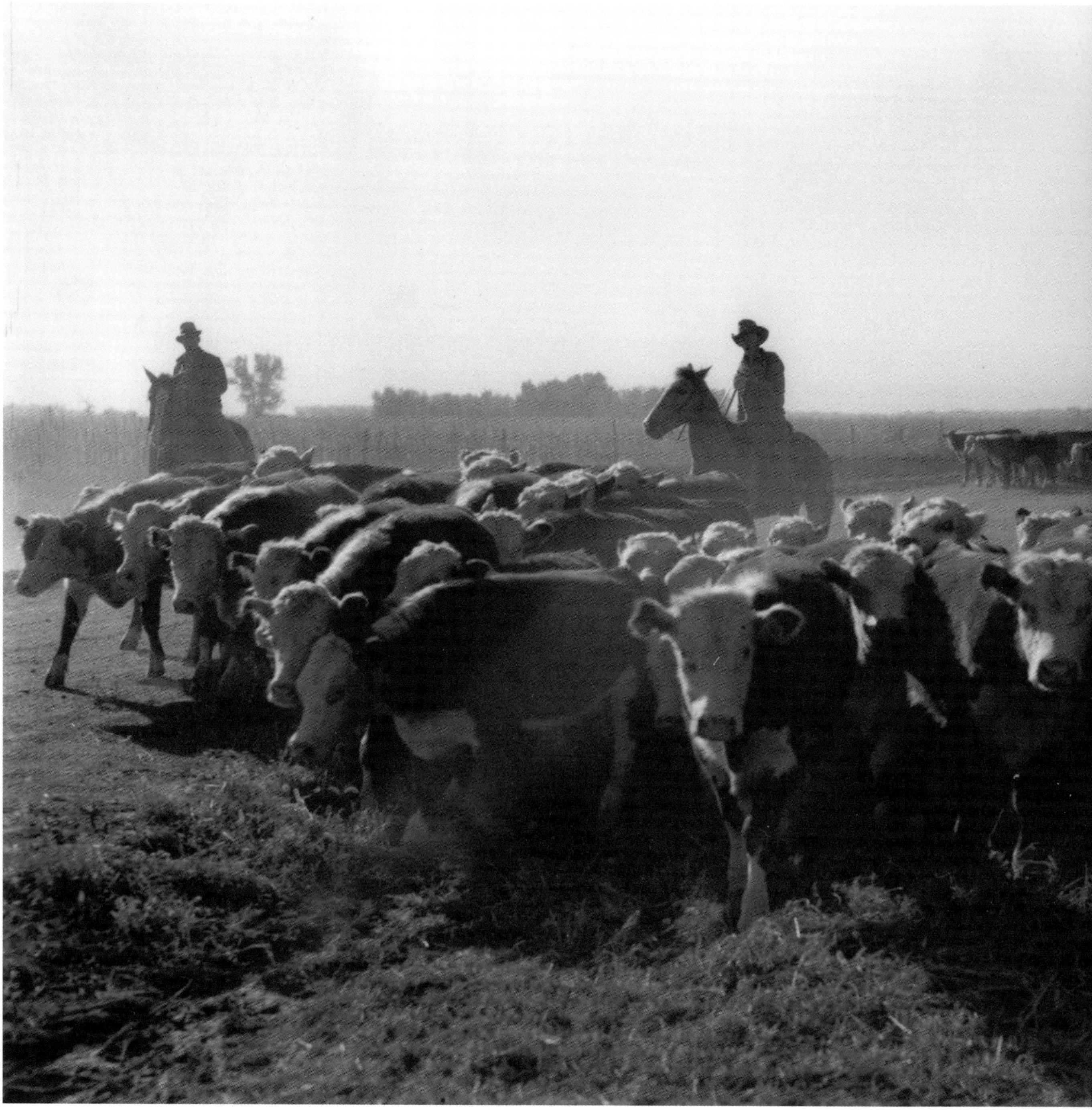

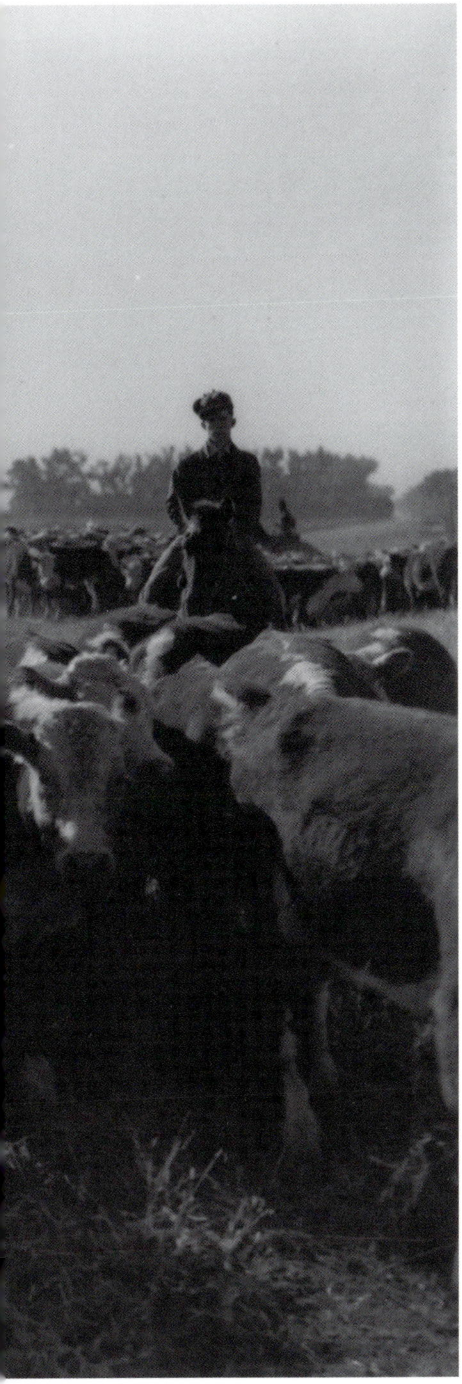

Cowboys and cattle. Dawson County, Nebraska. John Vachon, October 1938.
LC-USF34-008808-D DLC

Barn on Spokane Ranch, Big Hole Basin. Beaverhead County, Montana.
John Vachon, April 1942.
LC-USF34-065388-D DLC

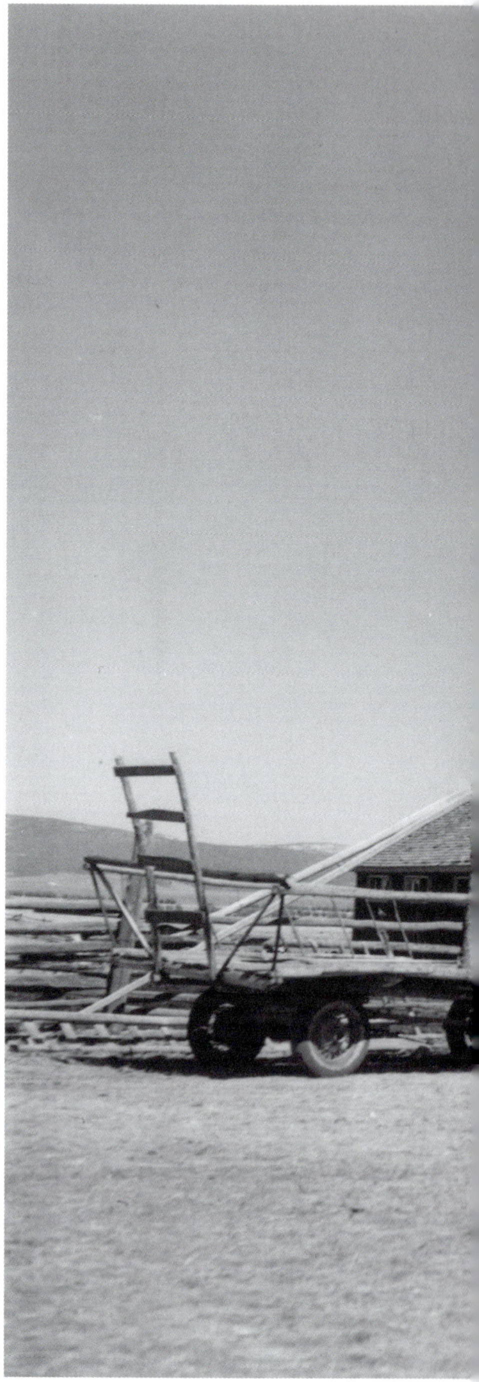

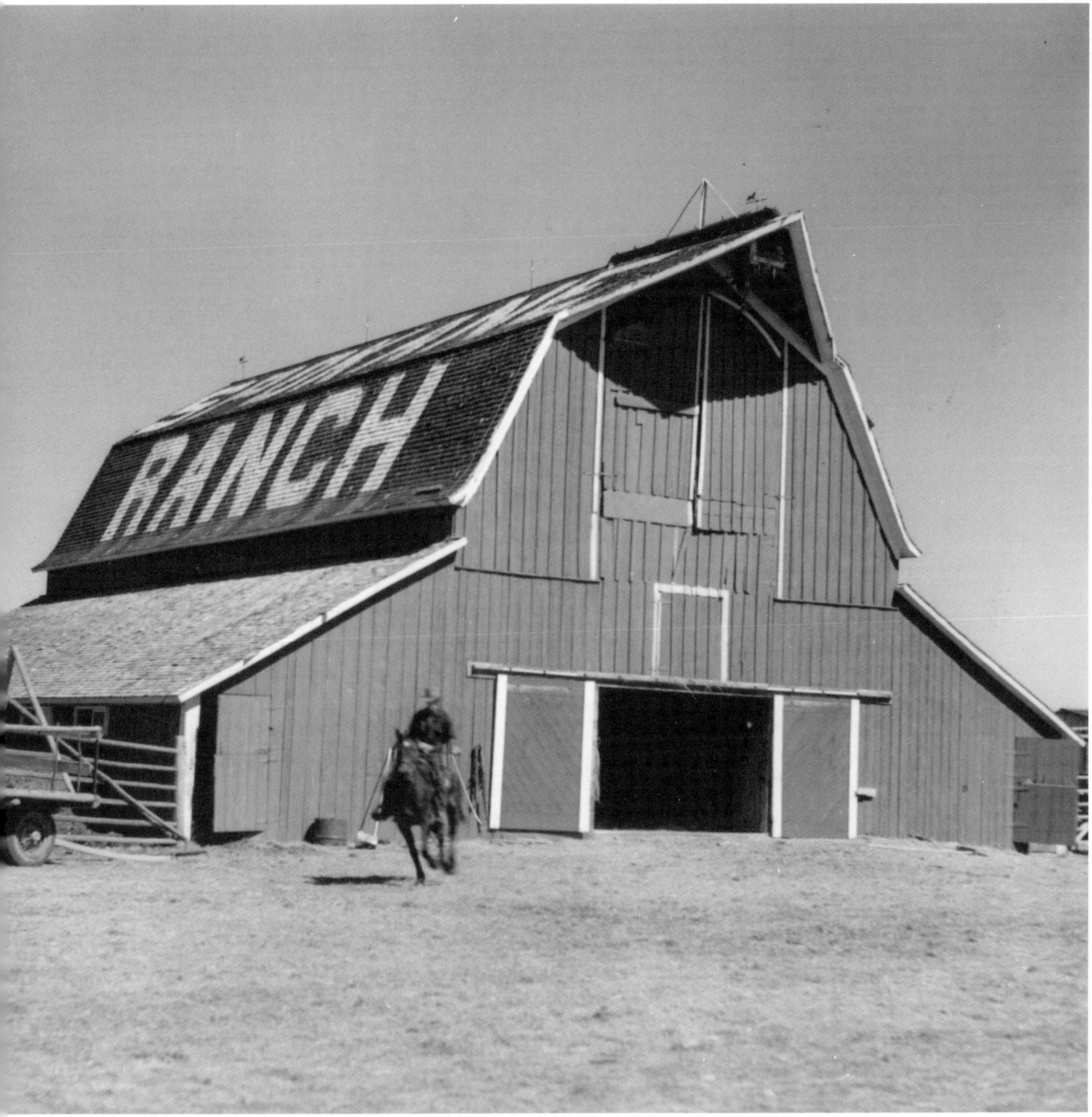
RANCH

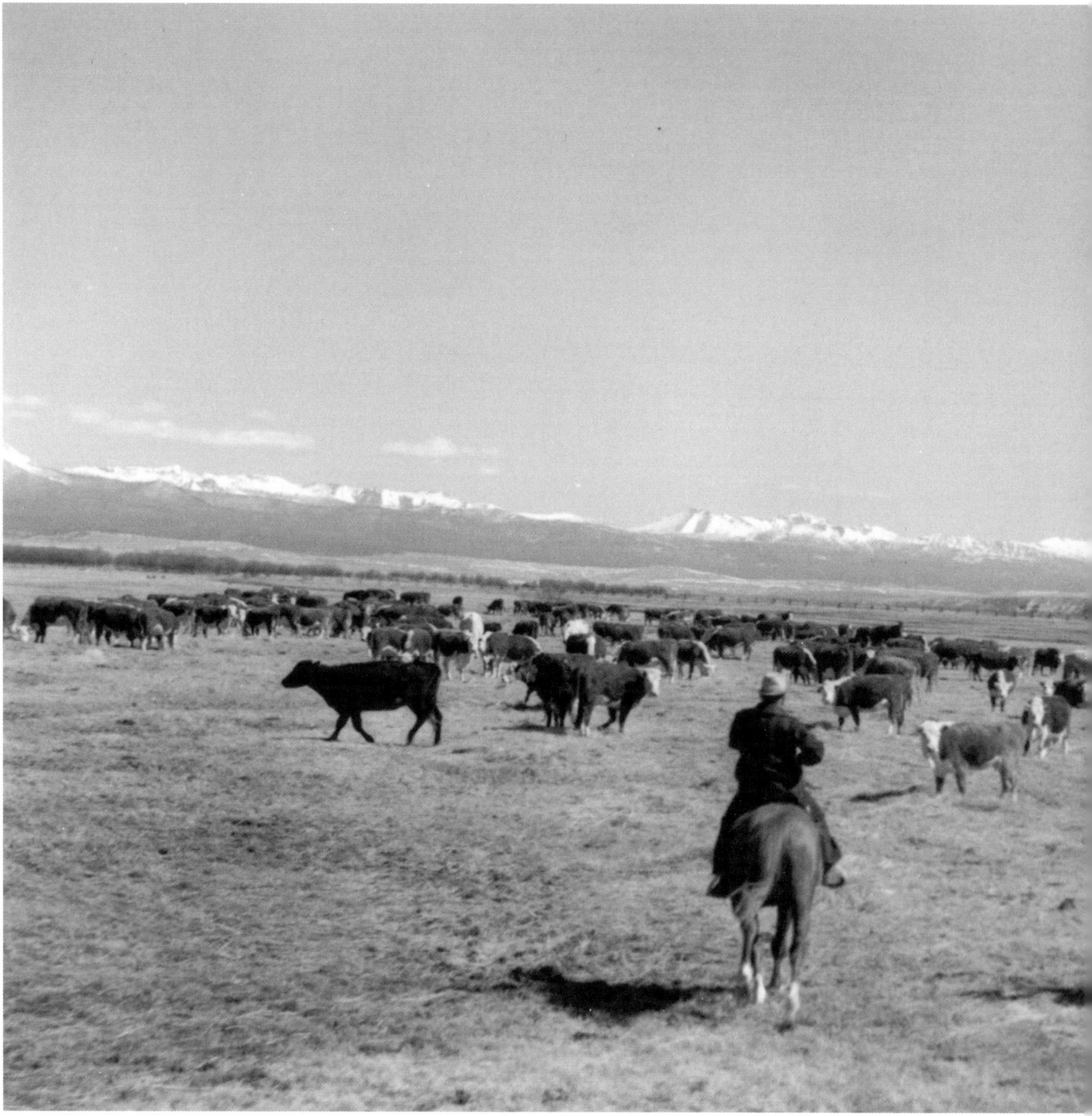

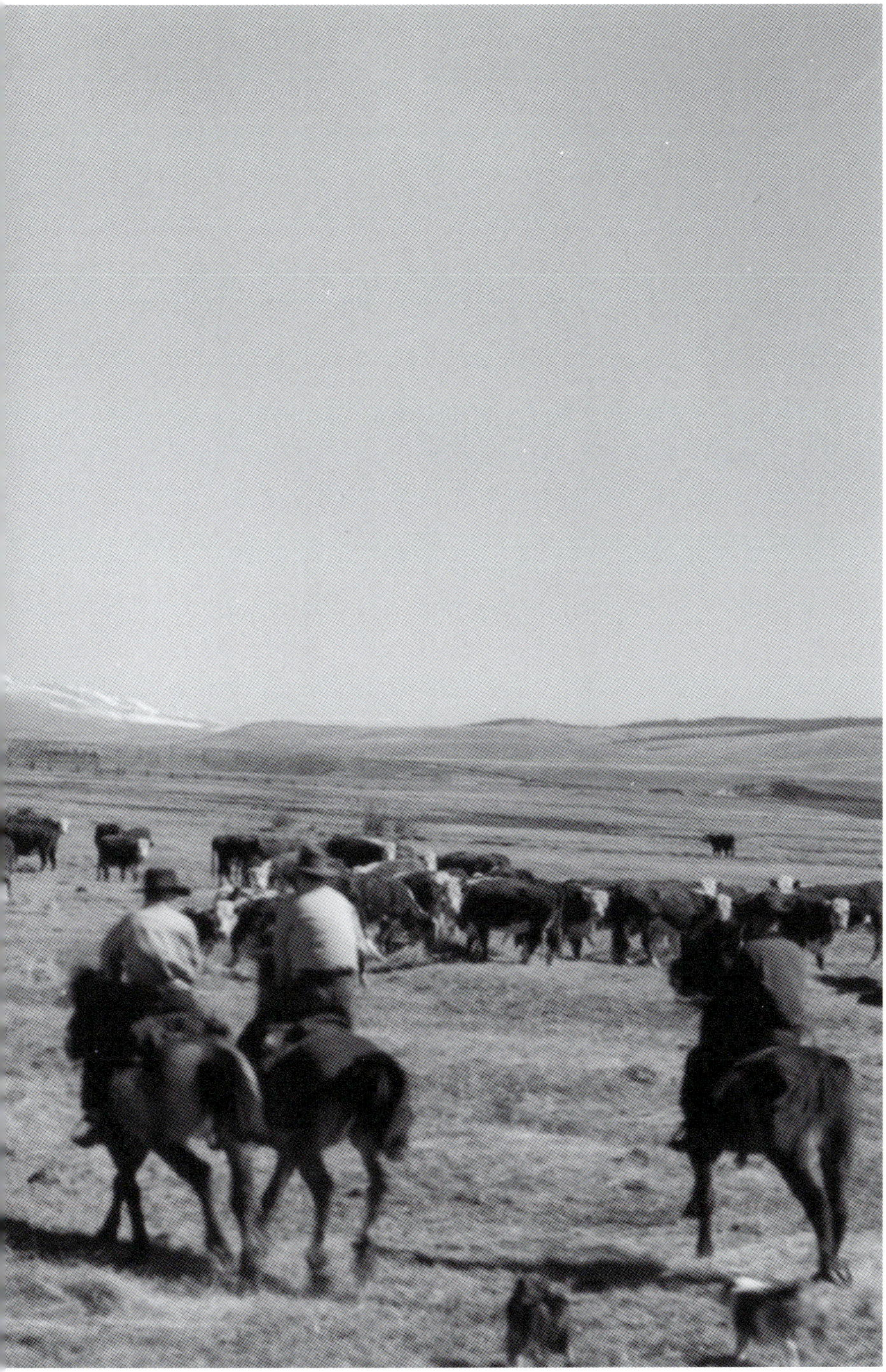

Hired hands riding back to the ranch after spreading a load of hay for cattle. Spokane Ranch, Beaverhead County, Montana. John Vachon, April 1942.
LC-USF34-065390-D DLC

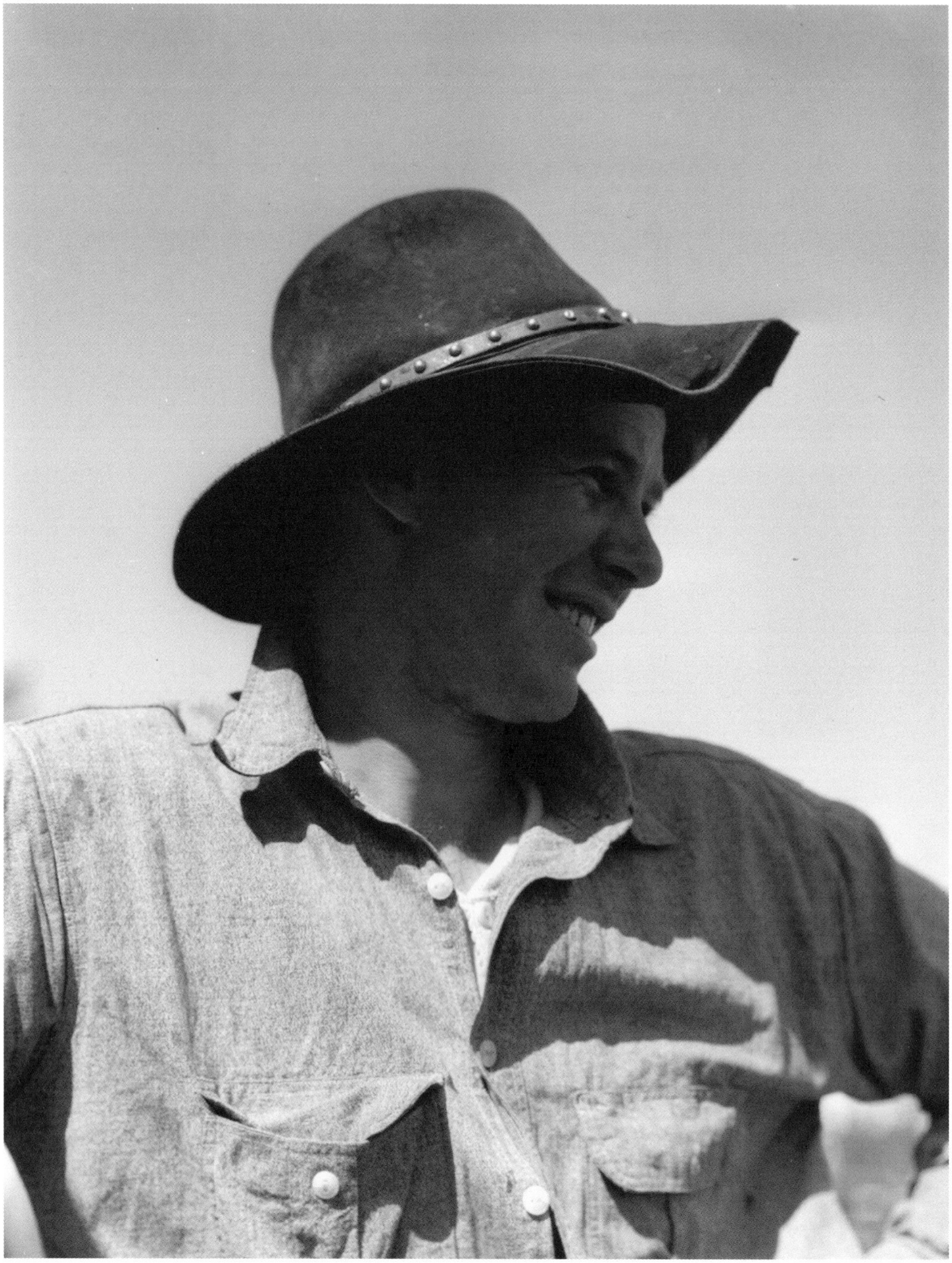

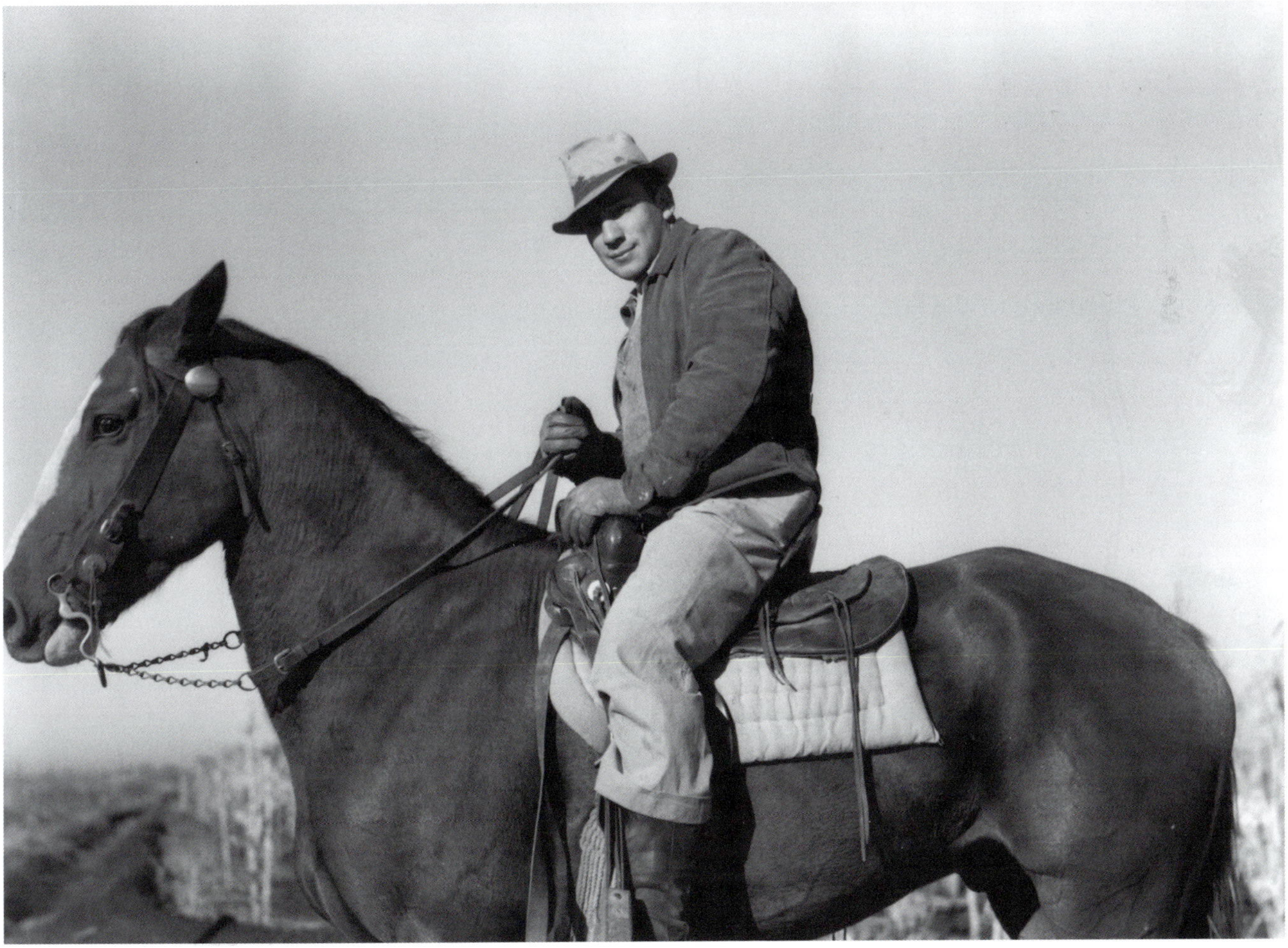

OPPOSITE: Cowboy. Bitterroot Valley, Ravalli County, Montana. John Vachon, April 1942. *LC-USF34-065300-D DLC*

ABOVE: Cowboy. Dawson County, Nebraska. John Vachon, October 1938. *LC-USF34-008811-D DLC*

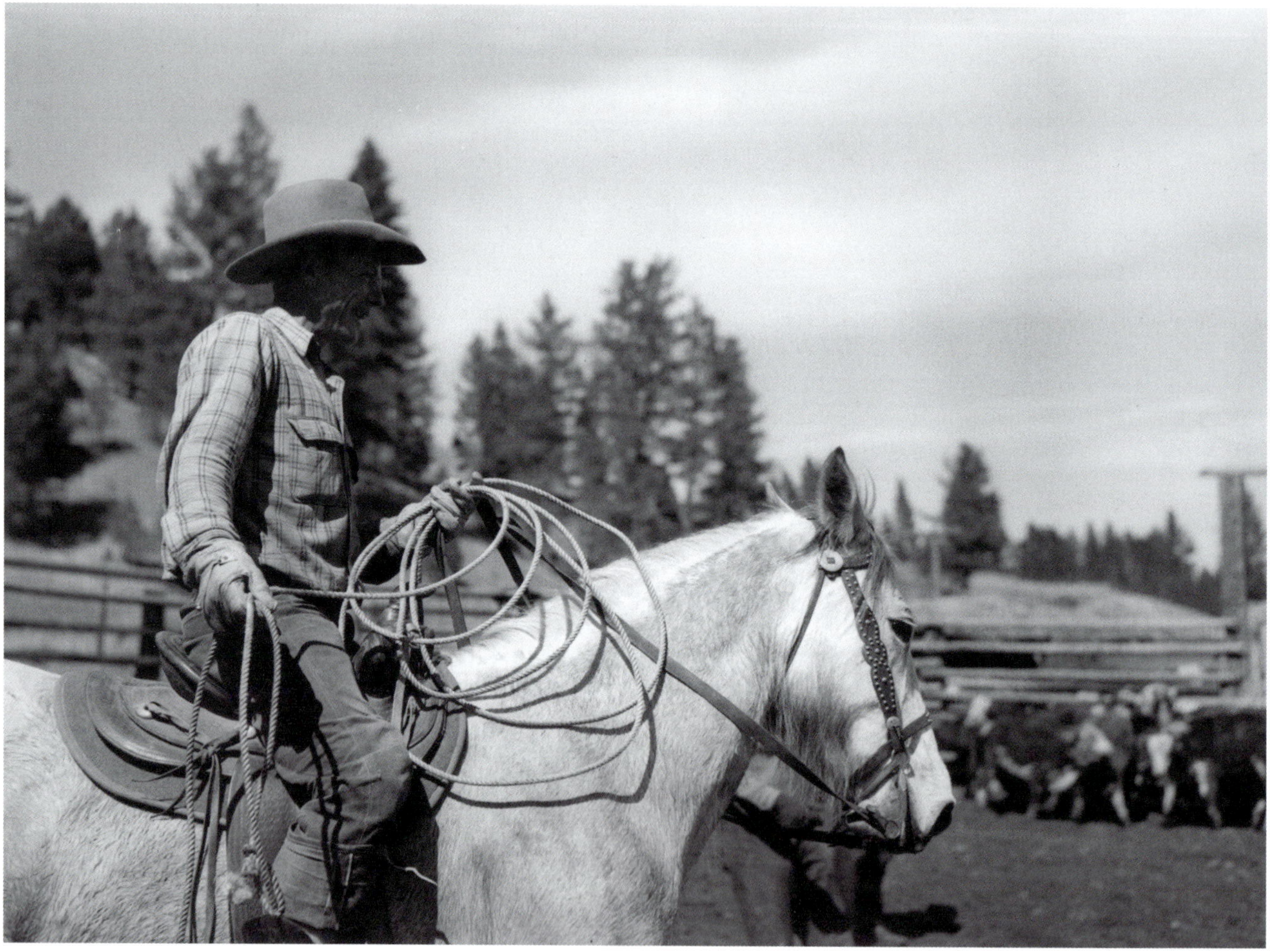

ABOVE: Cowboy. Bitterroot Valley, Ravalli County, Montana. John Vachon, April 1942.
LC-USF34-065259-D DLC

OPPOSITE: Walt Stewart in a card game on Saturday afternoon. Mr. Stewart is a rancher who owns 500 head of cattle. Wisdom, Montana. John Vachon, April 1942.
LC-USF34-065438-D DLC

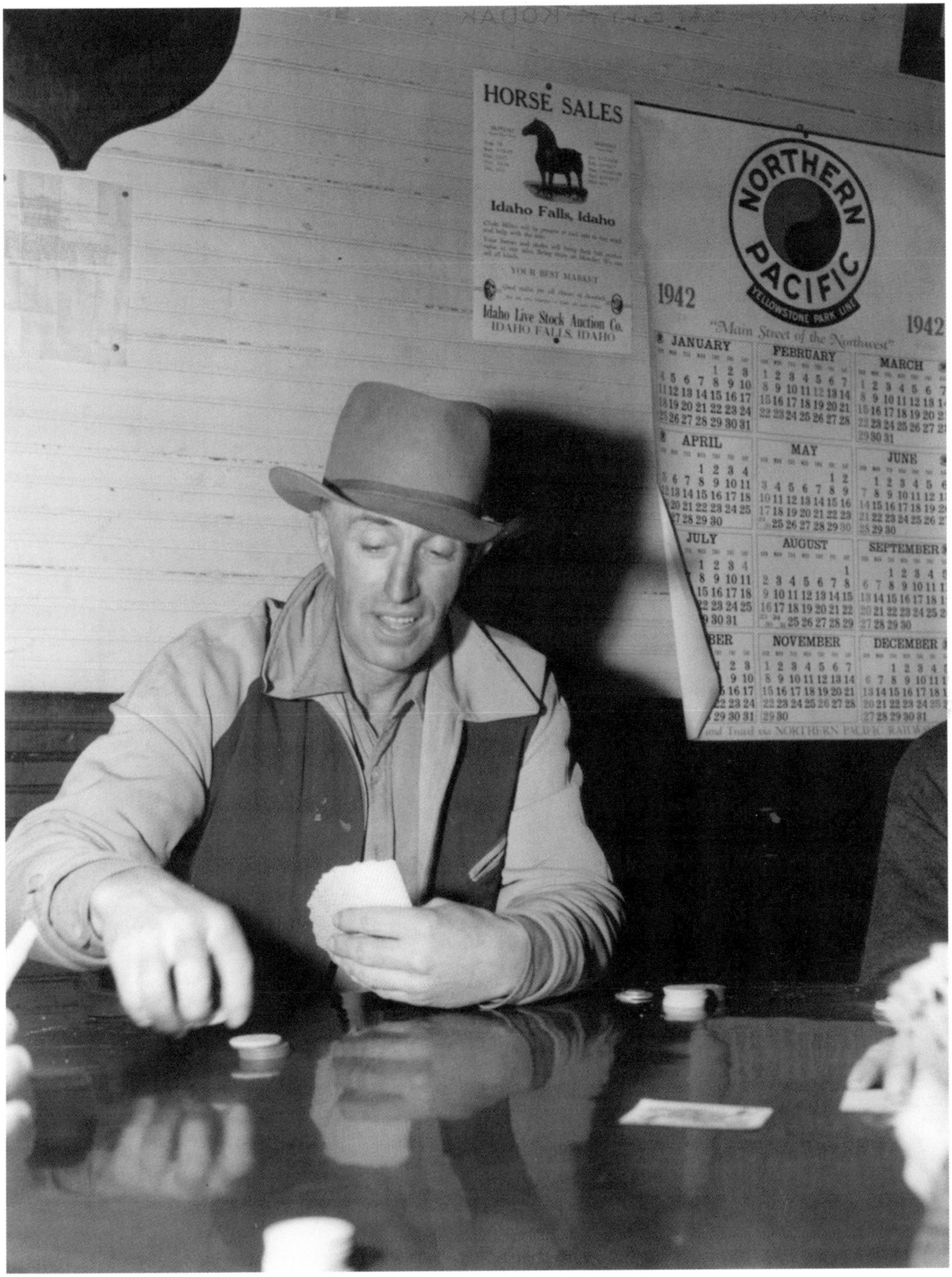
HORSE SALES
Idaho Falls, Idaho
YOUR BEST MARKET
Idaho Live Stock Auction Co.
IDAHO FALLS, IDAHO
NORTHERN PACIFIC
YELLOWSTONE PARK LINE
1942
1942
"Main Street of the Northwest"
JANUARY
FEBRUARY
MARCH
APRIL
MAY
JUNE
JULY
AUGUST
SEPTEMBER
NOVEMBER
DECEMBER

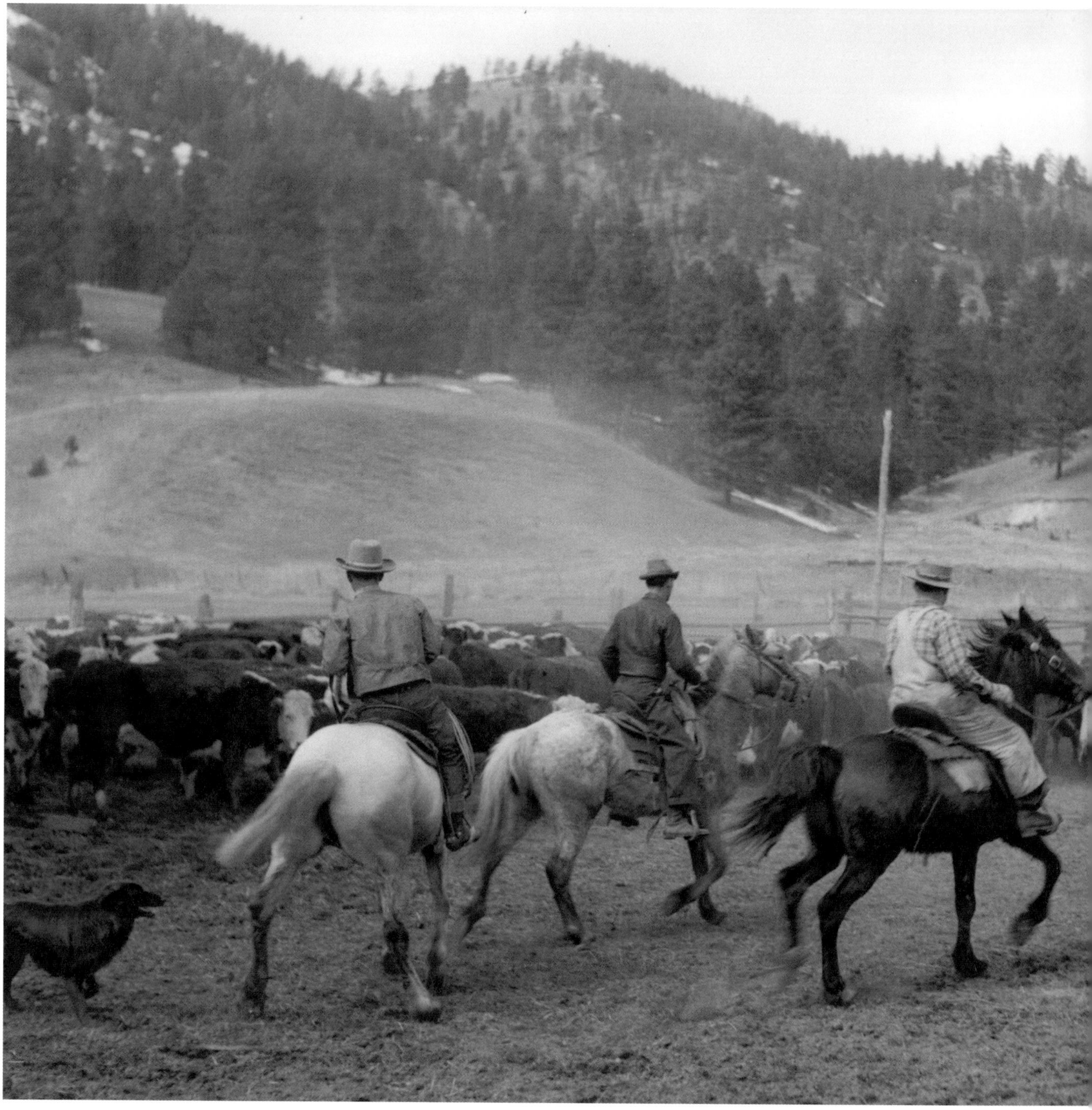

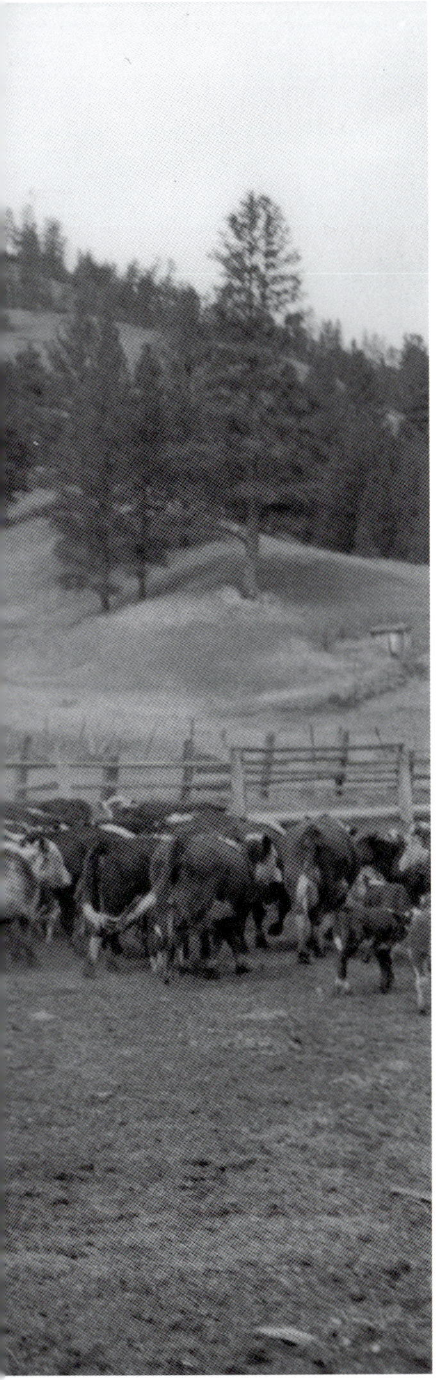

Driving cattle into the corral for branding and dehorning. Bitterroot Valley, Ravalli County, Montana. John Vachon, April 1942.
LC-USF34-065321-D DLC

ARTHUR ROTHSTEIN

1915–1985

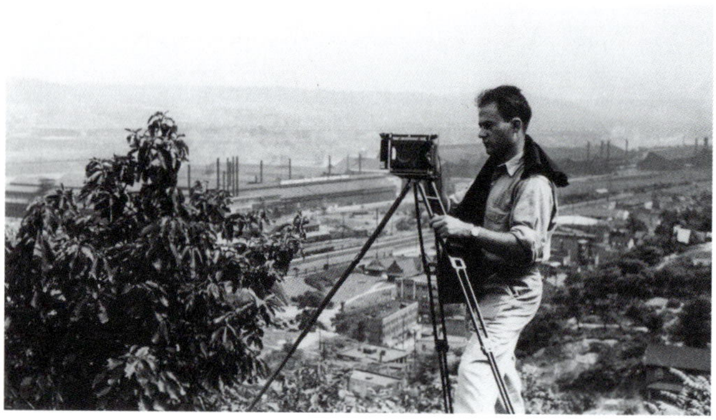

Arthur Rothstein, July 1938.
LC-DIG-ppmsc-00192

Arthur Rothstein was born in Manhattan, New York City, and grew up in the Bronx. He graduated from Columbia University in 1935, where he was a founder of the University Camera Club and photography editor of the *Columbian*. While at Columbia, he was fortunate to have Roy Stryker as one of his professors, and Rothstein served as Stryker's undergraduate assistant. Appreciating Rothstein's technical proficiency and enthusiasm for photography, Stryker hired him in 1935 as the first staff photographer for the Farm Security Administration, where he worked as a photographer for five years.

His first FSA assignment was to document the people of the Blue Ridge Mountains in Virginia and their "relocation." During his 50-year career as a photographer, Arthur Rothstein documented a great variety of subjects, including baseball games, war, struggling farmers, and U.S. presidents, but he was known primarily for his documentation of the Dust Bowl. His famous 1936 photograph of a farmer and his two sons in Cimarron County made the Oklahoma panhandle the visual symbol of the dust storms that devastated farmers in several Plains states. In 1936, he also photographed in Oregon and the Dakotas. Rothstein was one of four FSA photographers sent to Montana, along with John Vachon, Russell Lee, and Marion Post Wolcott, where he extensively photographed cowboy and ranch life, mostly in 1939. Of all the FSA photographers, Rothstein and Russell Lee were by far the most prolific photographers of ranching culture.

After leaving the FSA in 1940, Rothstein took a position as a photographer for *Look* magazine, where he remained until 1971, ultimately serving as the magazine's director of photography. He authored seven books on photojournalism featuring his photography.

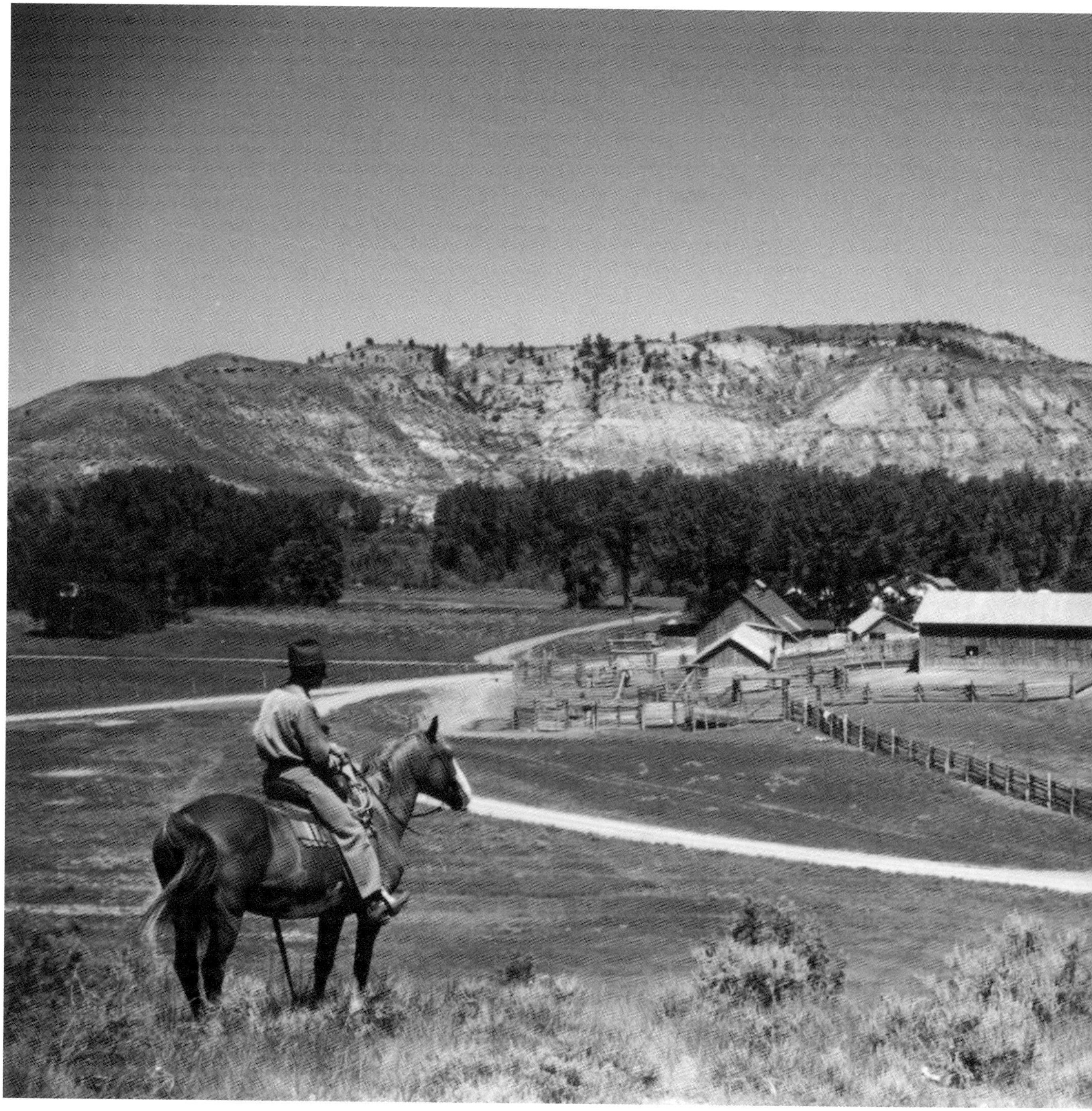

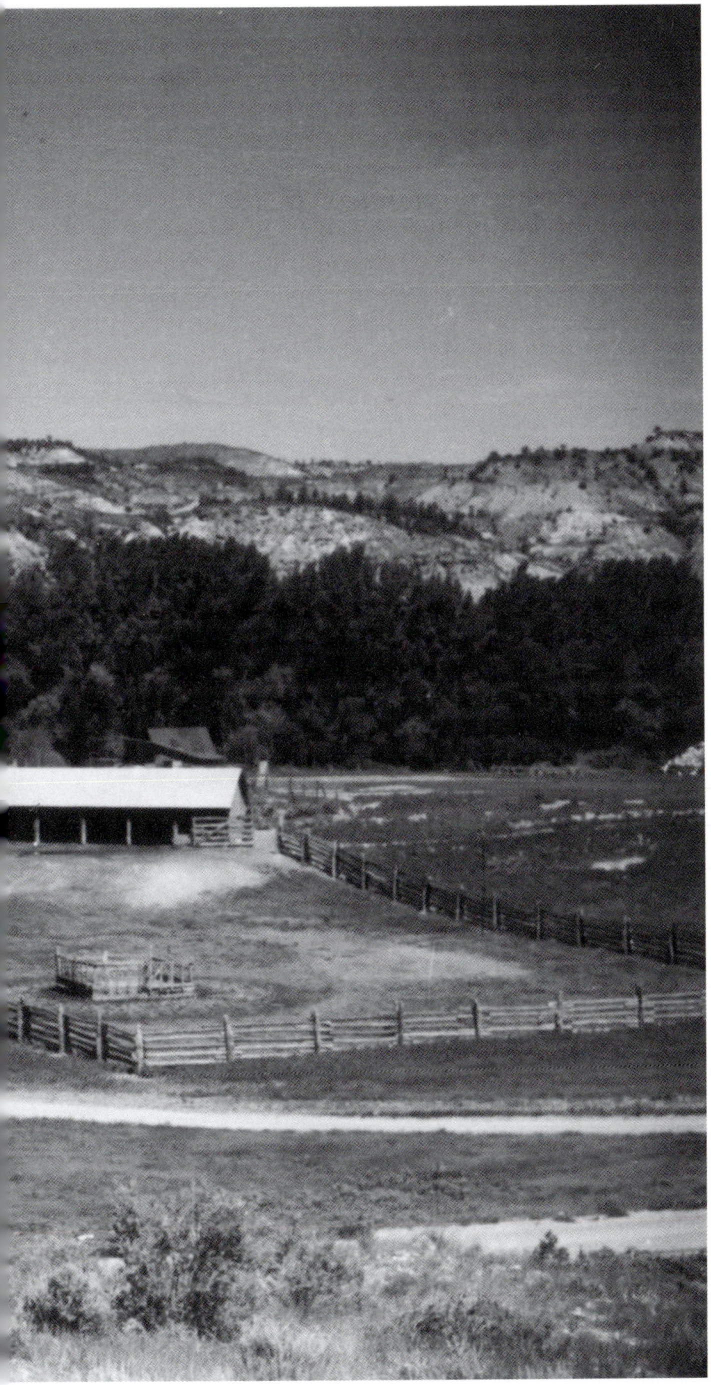

Jack Arnold looking over the Quarter Circle U Ranch, which he owns. Rosebud County, Montana. Arthur Rothstein, June 1939.
LC-USF33-003205-M1 DLC

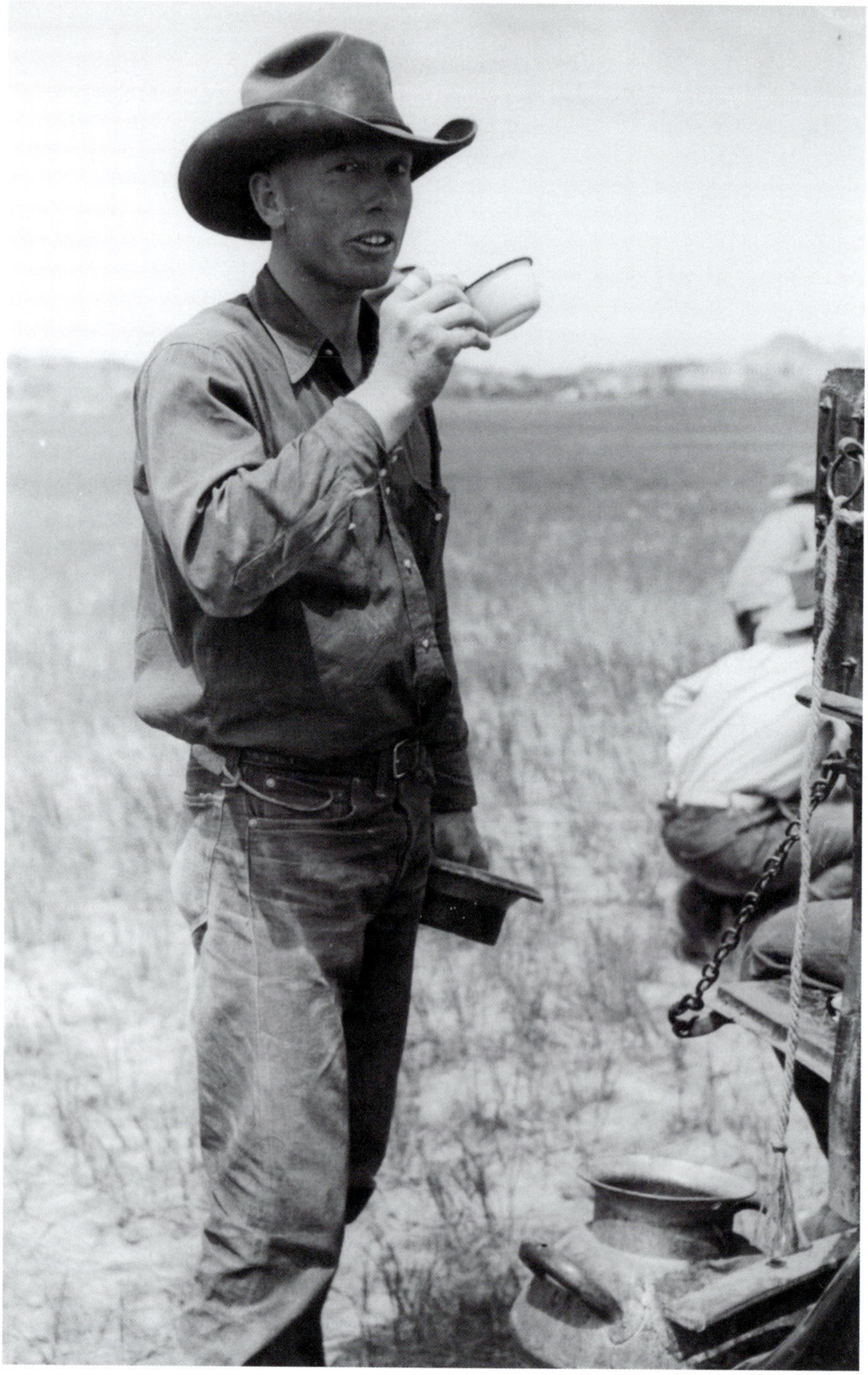

LEFT: Cowboy at Quarter Circle U Ranch roundup, Montana. Arthur Rothstein, June 1939. *LC-USF33-003295-M4 DLC*

OPPOSITE: Frank Latta, oldtime cowpuncher. Bozeman, Montana. Arthur Rothstein, June–July 1939. *LC-USF34-027197-D DLC*

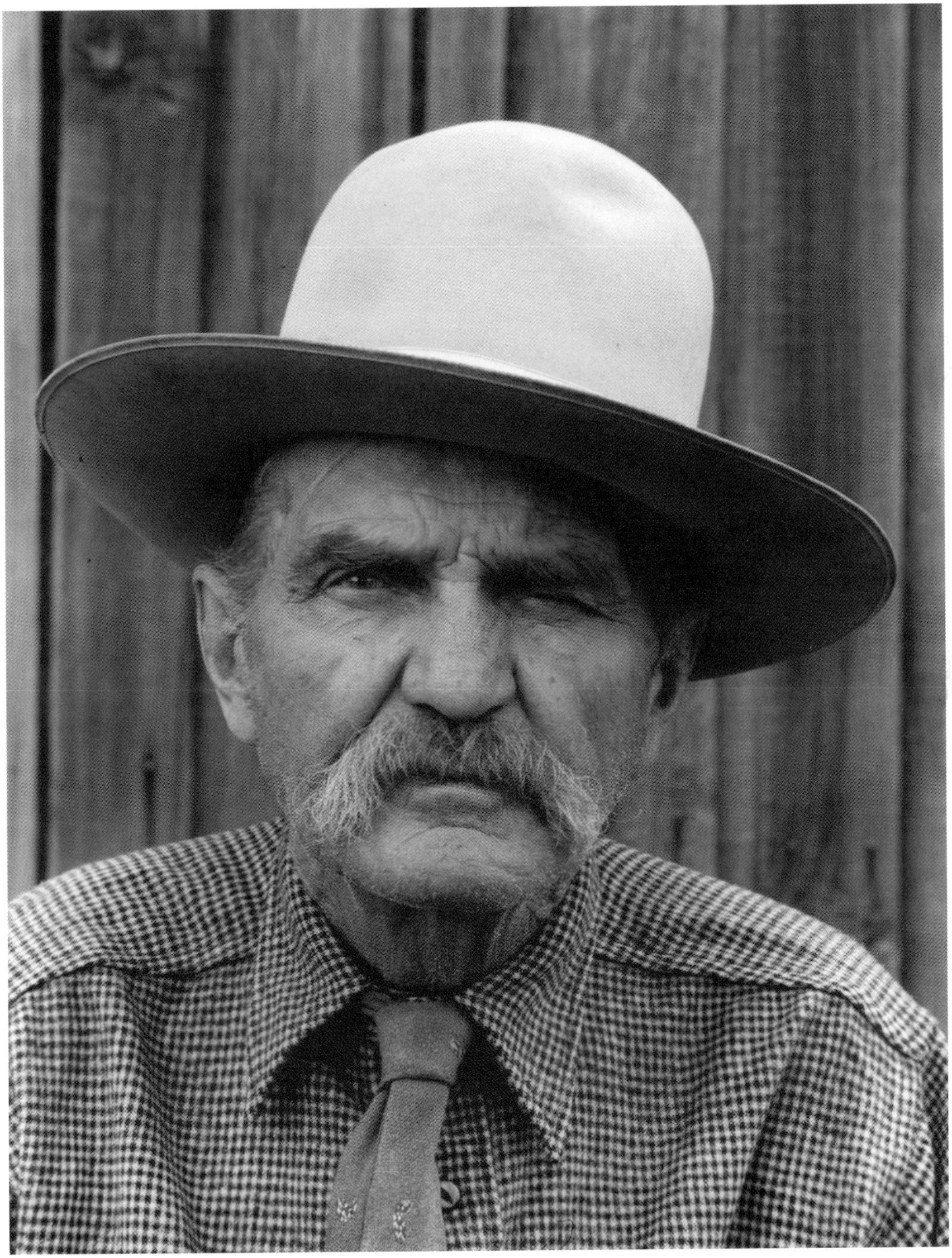

PICKUPS

The pickup truck is taken for granted as a mainstay on ranches today, but it was a godsend to ranchers when it first arrived. Ford's first factory-assembled pickup was based on the Ford Model T car, with a modified rear body. It debuted in 1925 and sold for $281. Henry Ford billed it as the "Ford Model T Runabout with Pickup Body." The 34,000 pickups built that first year featured a cargo box, adjustable tailgate, four stake pockets, and heavy-duty rear springs. In 1928, the Model A replaced the Model T, introducing the first closed-cab pickup. It sported innovations such as a safety glass windshield, roll-up side windows, and a three-speed transmission. It was powered by a four-cylinder L-head engine capable of 40 horsepower. 1918 was the first year for a factory-built Chevrolet pickup, known as the "Independence Series." In 1932, the 65-horsepower Ford flathead V8 engine was offered as an option in the truck. By 1936, Ford had already produced 3 million trucks and led the industry in sales.

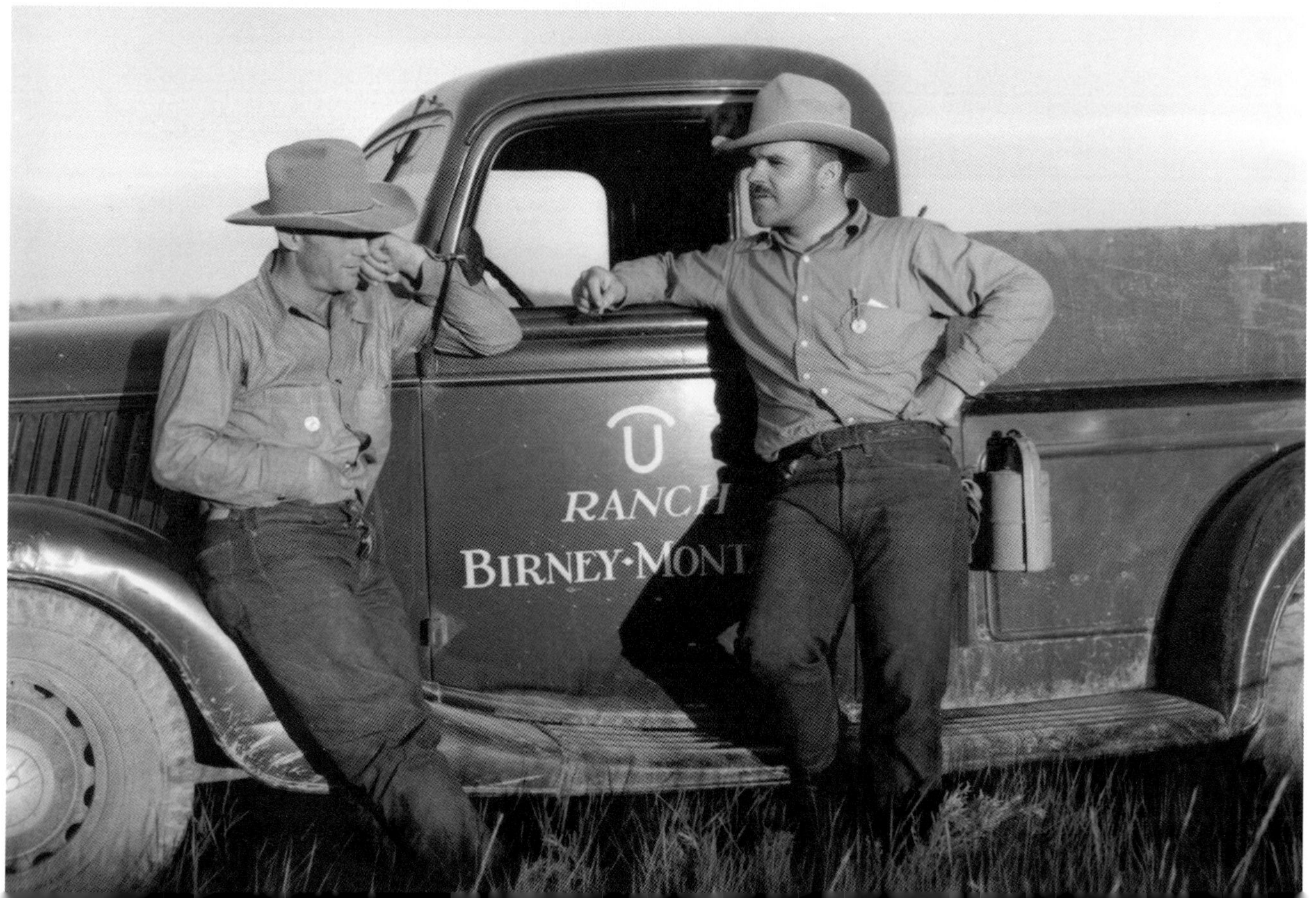

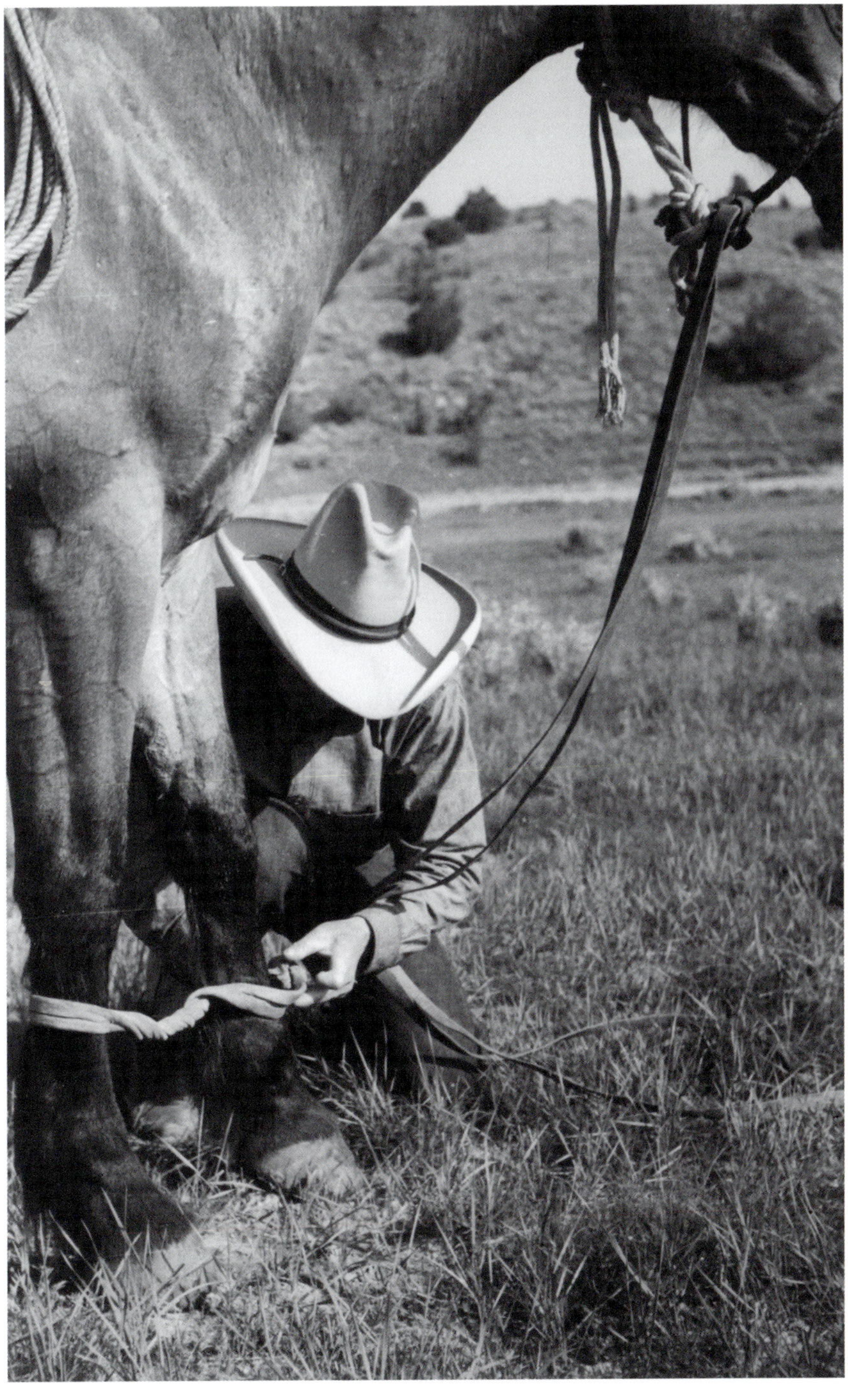

OPPOSITE: Rancher and cowhand. Quarter Circle U Ranch. Big Horn County, Montana. Arthur Rothstein, June 1939.
LC-USF34-027829-D DLC

RIGHT: Putting a hobble on a horse. Quarter Circle U Brewster-Arnold Ranch Company near Birney, Montana. Arthur Rothstein, June 1939.
LC-USF33-003138-M3 DLC

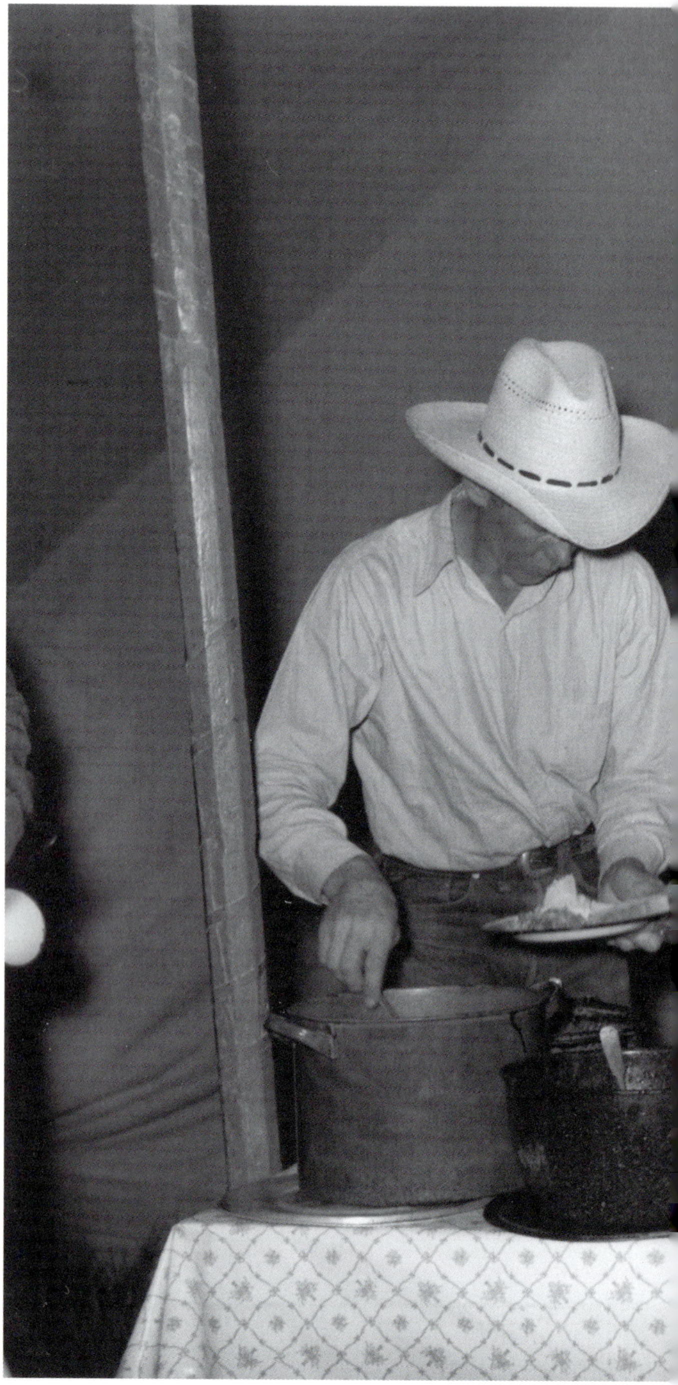

Cowhands at dinner during the roundup. Quarter Circle U Brewster-Arnold Ranch Company near Birney, Montana. Arthur Rothstein, June 1939. *LC-USF34-027710-D*

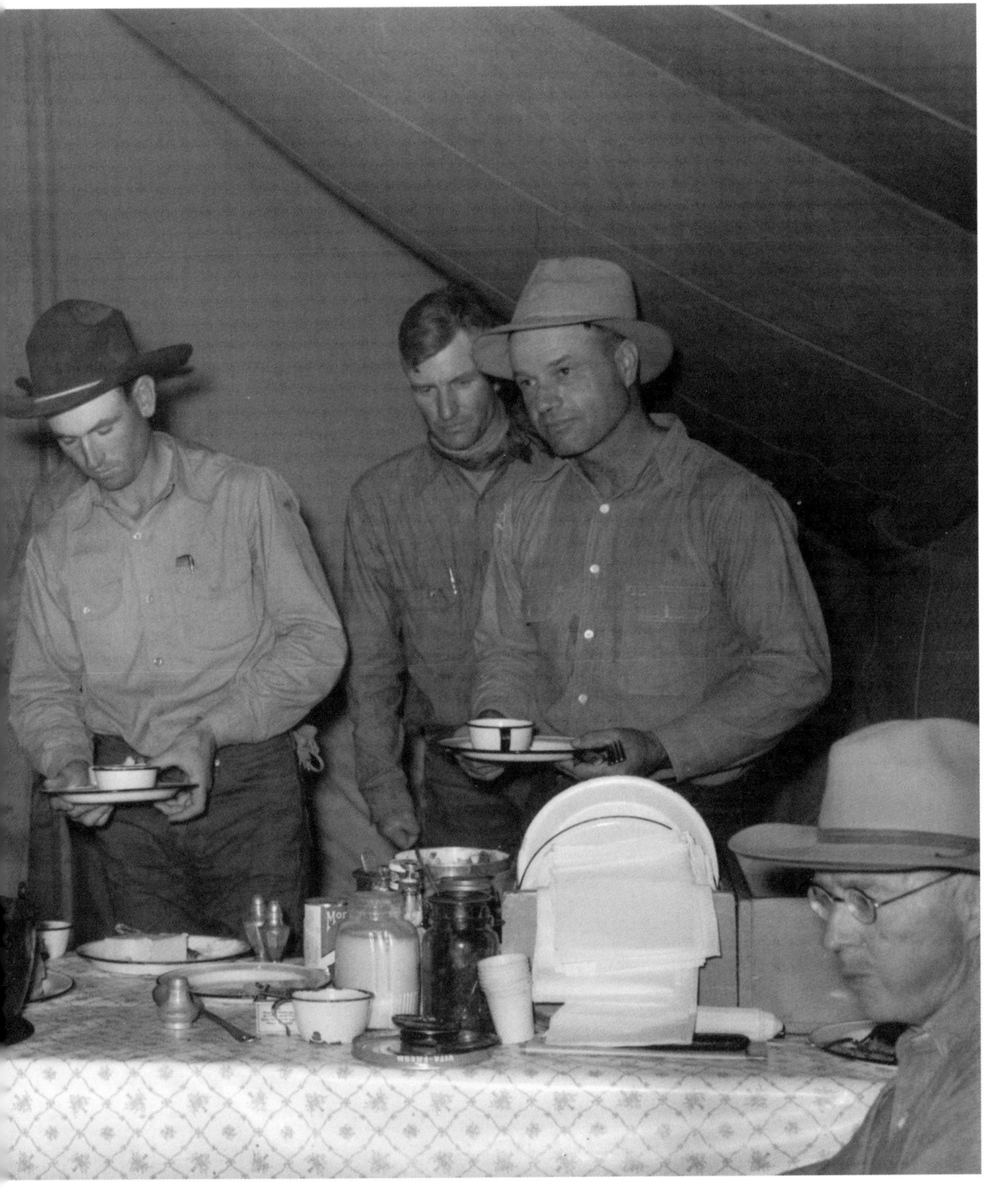

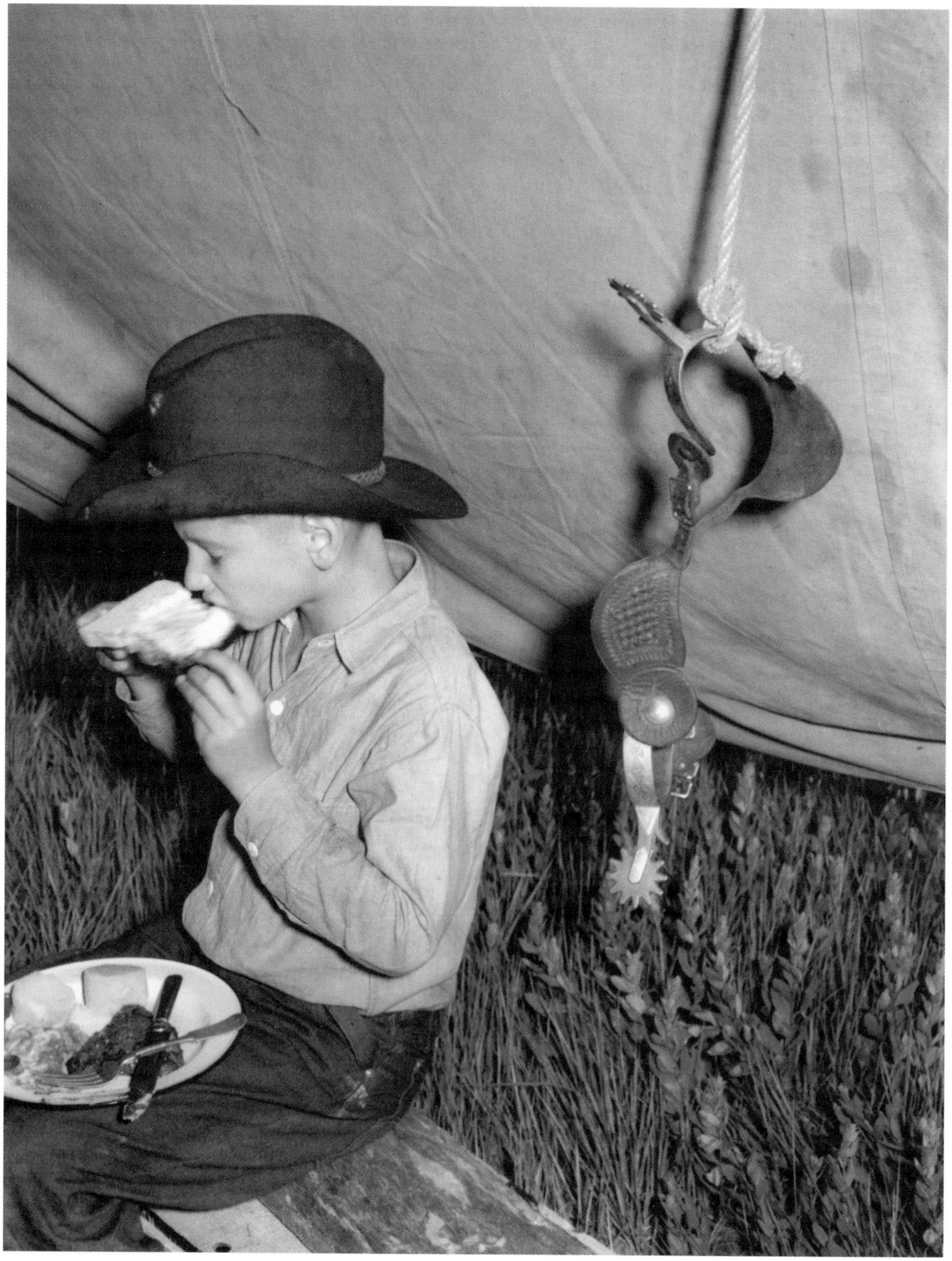

 Arthur Rothstein

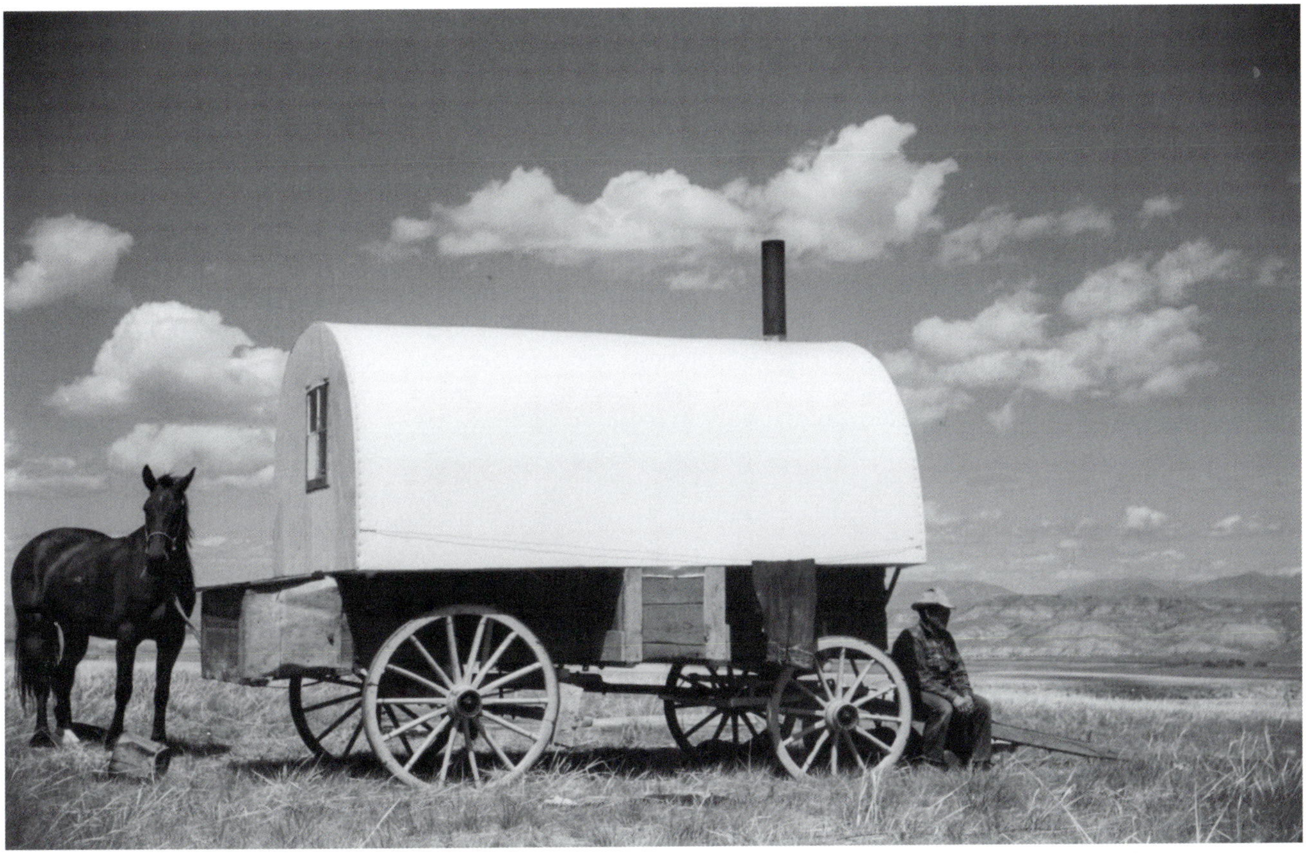

OPPOSITE: Rancher's son at dinner. Quarter Circle U Ranch roundup. Big Horn County, Montana. Arthur Rothstein, June 1939.
LC-USF34-027723

ABOVE: Sheepherder and chuck wagon [i.e., sheepherder's wagon]. Madison County, Montana. Arthur Rothstein, Summer 1939.
LC-USF33-003080-M4 DLC

COWBOY MUSIC

Cowboy songs, essentially cowboy poetry set to music, were an important part of cowboy culture from the earliest trail-driving days after the Civil War. Cowboy songs are occupational folk songs about the life and work of the working cowboy. Many cowboy songs were based on much older songs and ballads from the British Isles. "The Streets of Laredo" was a cowboy re-working of the old Irish ballad "The Unfortunate Rake" about a soldier dying of syphilis. "Bury Me Not on the Lone Prairie" was a reworking of an older sailor's song "The Ocean Burial (Bury Me Not in the Deep, Deep Sea)" from the 1830s. Many of our best-known cowboy songs, however, were made by working cowboys with a knack for versifying, like Montanan D.J. O'Malley, who wrote "When the Work's All Done This Fall," or Arizona cowboy Gail Gardner, who wrote "Tying Knots in the Devil's Tail." These songs would go into oral tradition, and in the early 1900s began to appear in print as well. In isolated cow camps or ranch bunkhouses, singing songs, reciting poetry, and telling stories were welcome forms of entertainment.

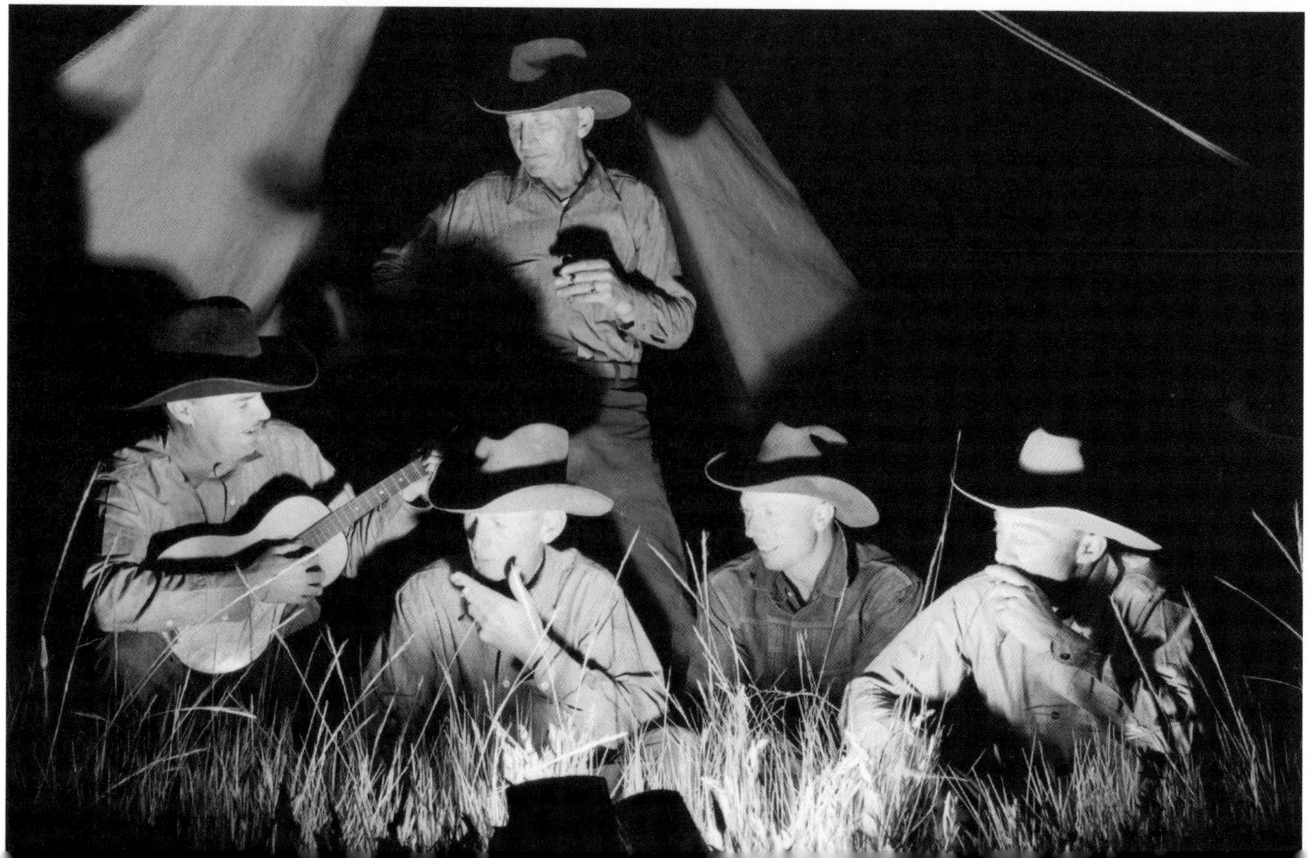

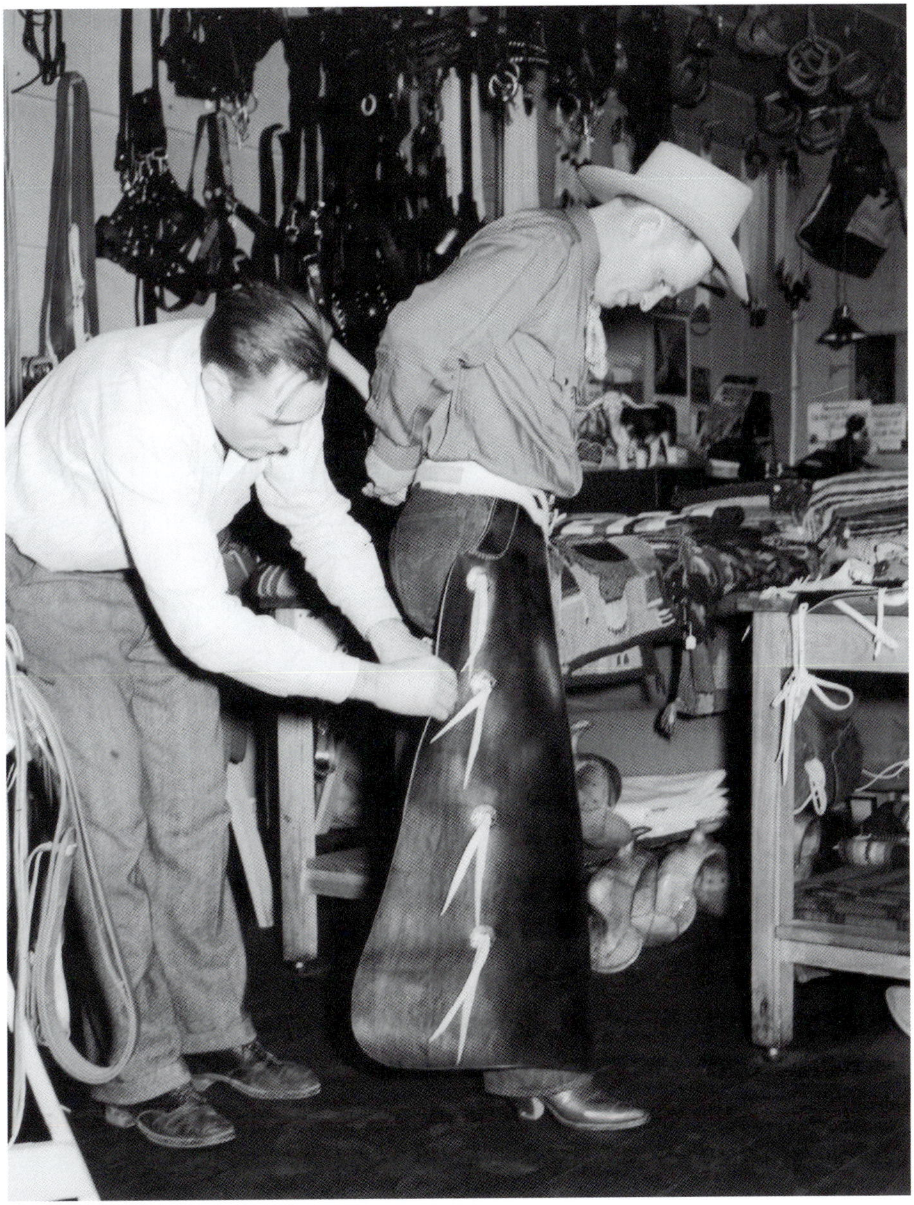

OPPOSITE: Cowhands singing after the day's work. Quarter Circle U Ranch roundup. Big Horn County, Montana. Arthur Rothstein, June 1939.
LC-USF346-027824-D DLC

RIGHT: Cowhand tries on chaps at the Capriola Saddlery. Elko, Nevada. Arthur Rothstein, March 1940.
LC-USF34-029948-D DLC

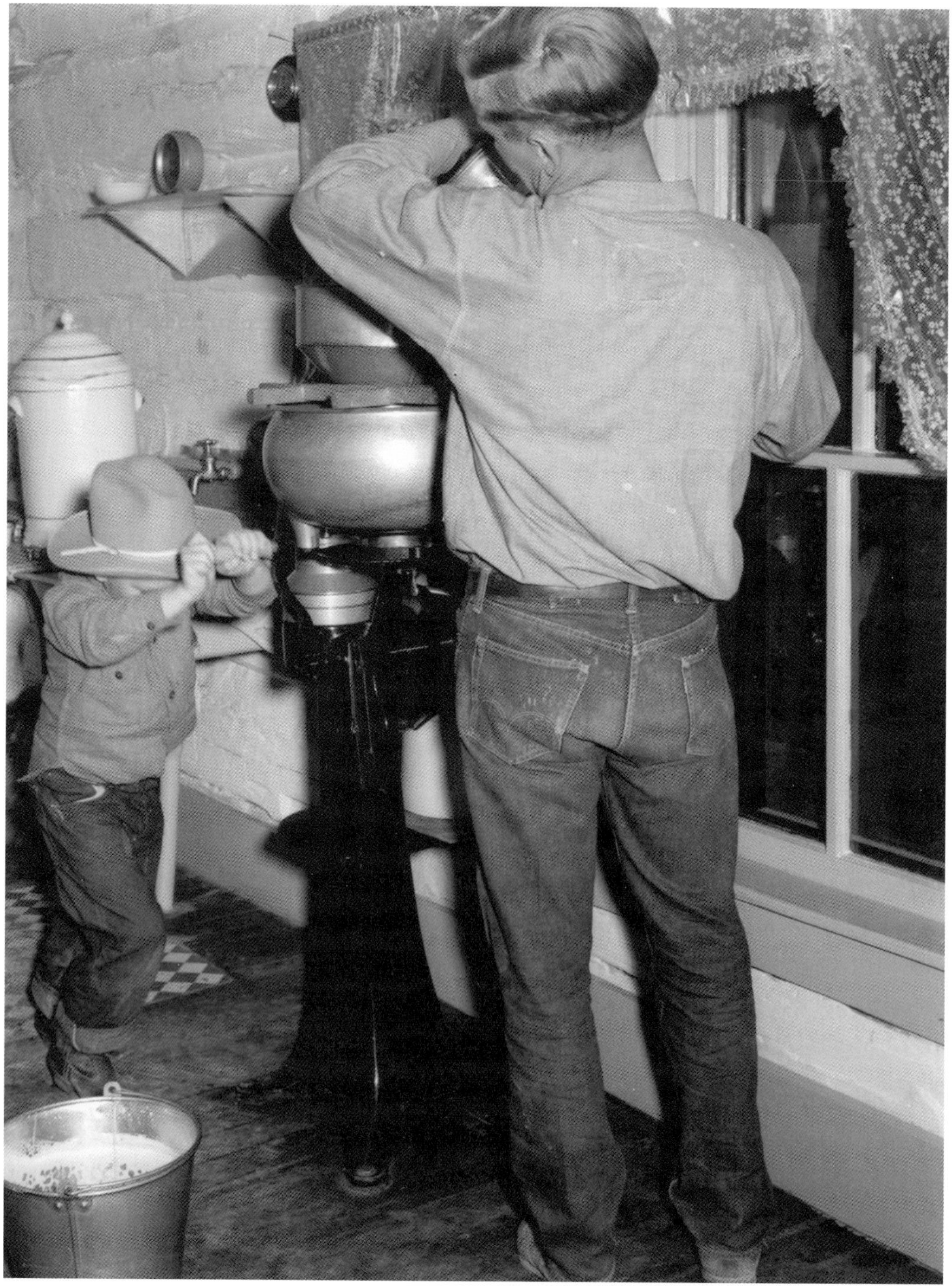

OPPOSITE: Running a cream separator. Quarter Circle U Ranch, Montana. Arthur Rothstein, June 1939. *LC-USF34-027647-D DLC*

RIGHT: Cowboy in front of the bunkhouse, Quarter Circle U Ranch. Big Horn County, Montana. Arthur Rothstein, June 1939. *LC-USF33-003169-M4 DLC*

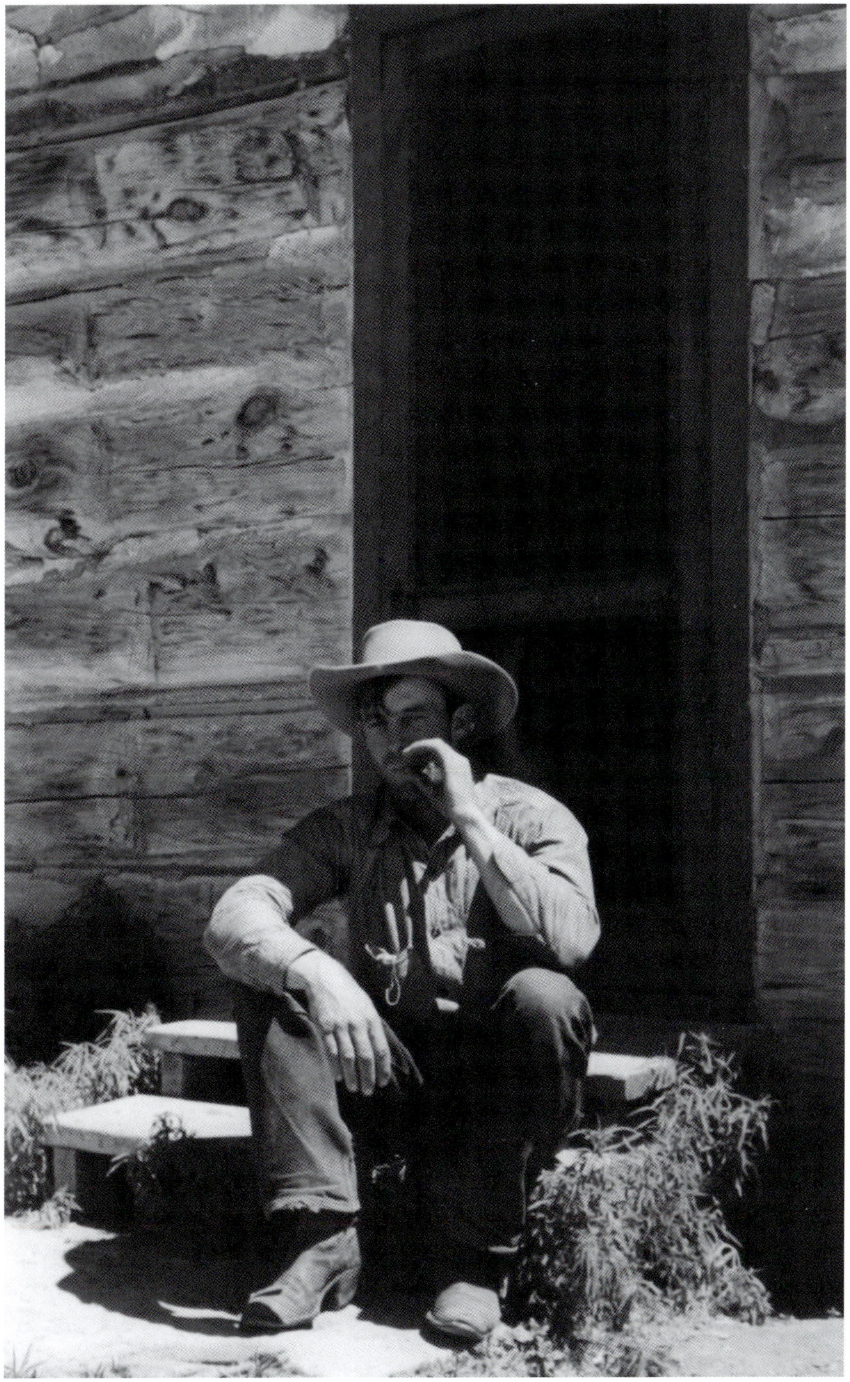

TELEPHONE

Rural ranches were often quite isolated, and the only way to see one's neighbors was to go visit them. Getting news of the outside world might involve a trip to town. The first telephones came to rural areas around the turn of the last century. In 1902, few ranches or farms had telephones, but by 1920 thirty-nine percent had them. The number of rural phone users dropped during the Depression years when only about twenty-five percent of ranches and farms had them. The telephone provided communication with neighbors and also stores and services in town. Ranchers could get weather information on the phone. Help could be summoned in case of danger or an emergency. And, of course, the early party lines created a new and much-used conduit for local gossip.

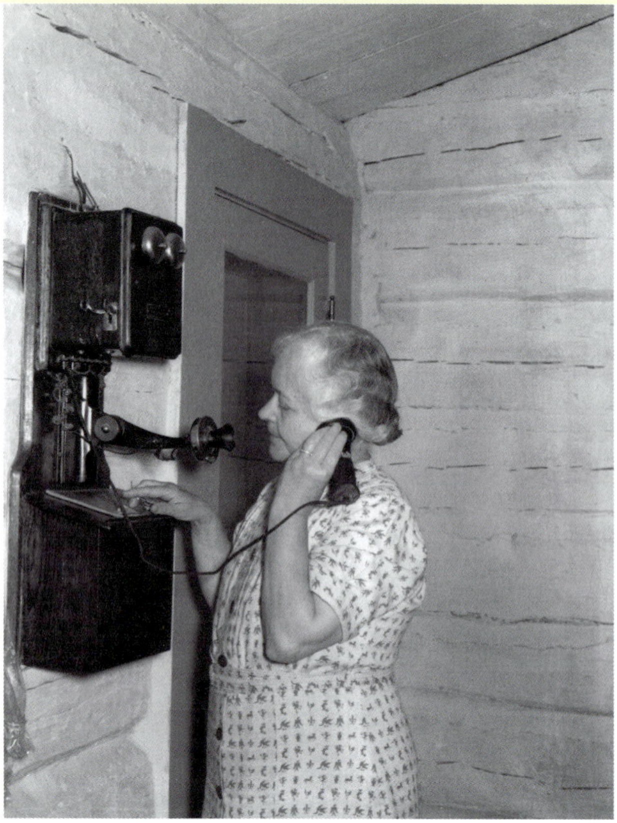

LEFT: Wife of ranch owner on party line phone. Quarter Circle U Ranch. Big Horn County, Montana. Arthur Rothstein, June 1939.
LC-USF34-027449-D

OPPOSITE: Girl with pony. Warren Brewster Ranch, Montana. Arthur Rothstein, June 1939.
LC-USF34-027640-D DLC

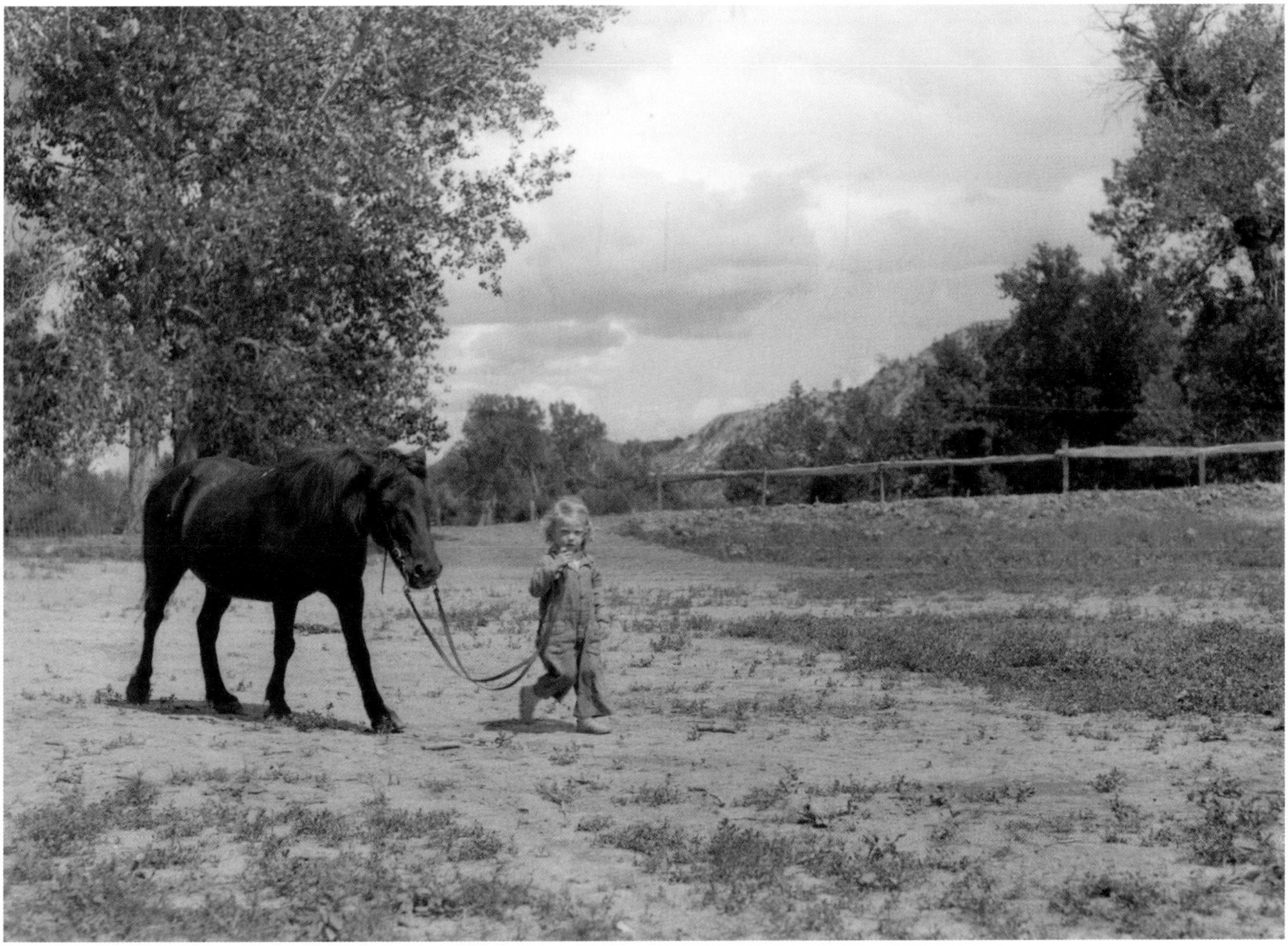

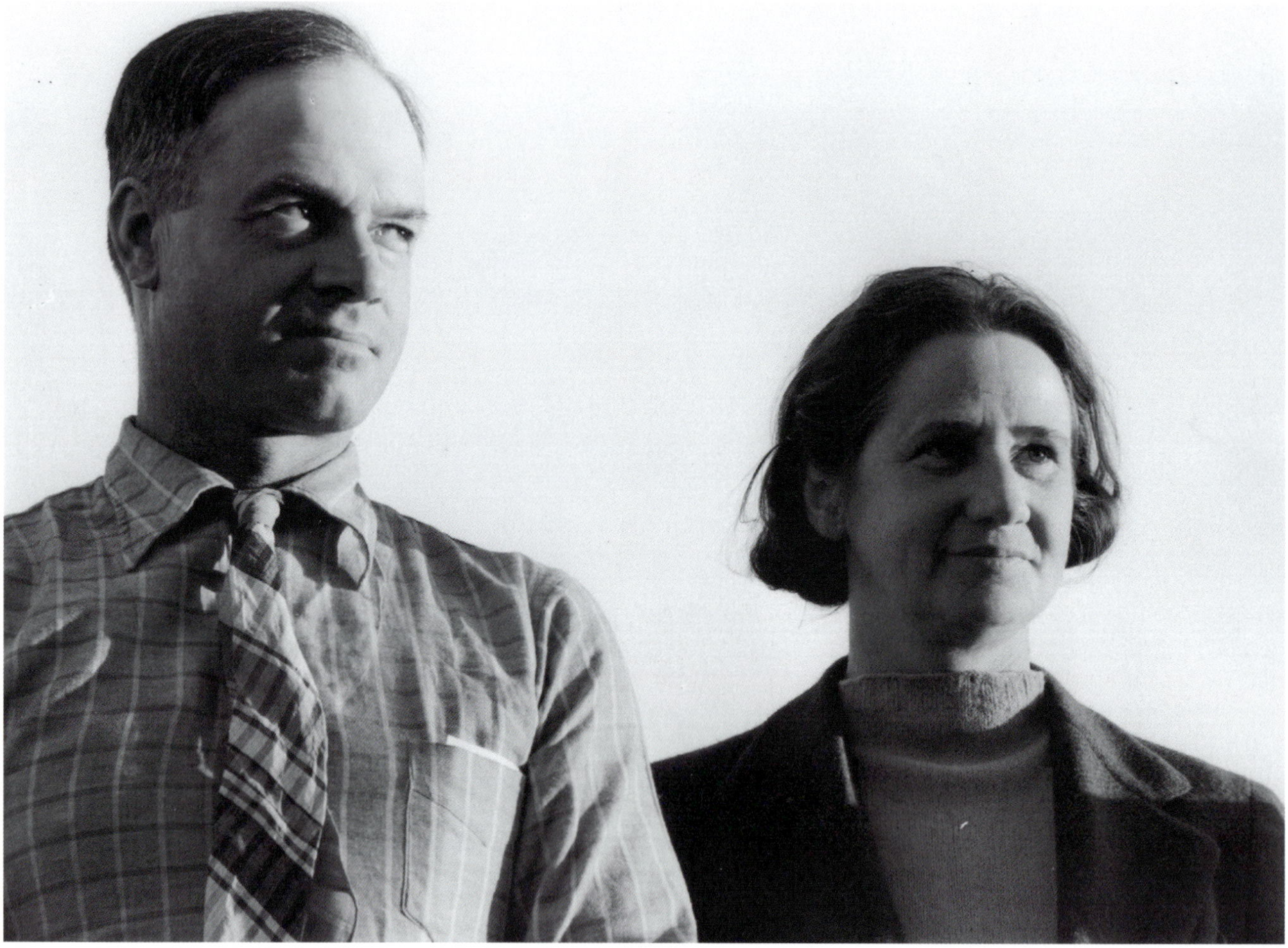

ABOVE: Warren Brewster and wife, ranchers. Warren Brewster Ranch, Montana. Arthur Rothstein, June 1939. *LC-USF34-027589-D*

OPPOSITE: Cowhand. Elko County, Nevada. Arthur Rothstein, March 1940. *LC-USF34-029966-D DLC*

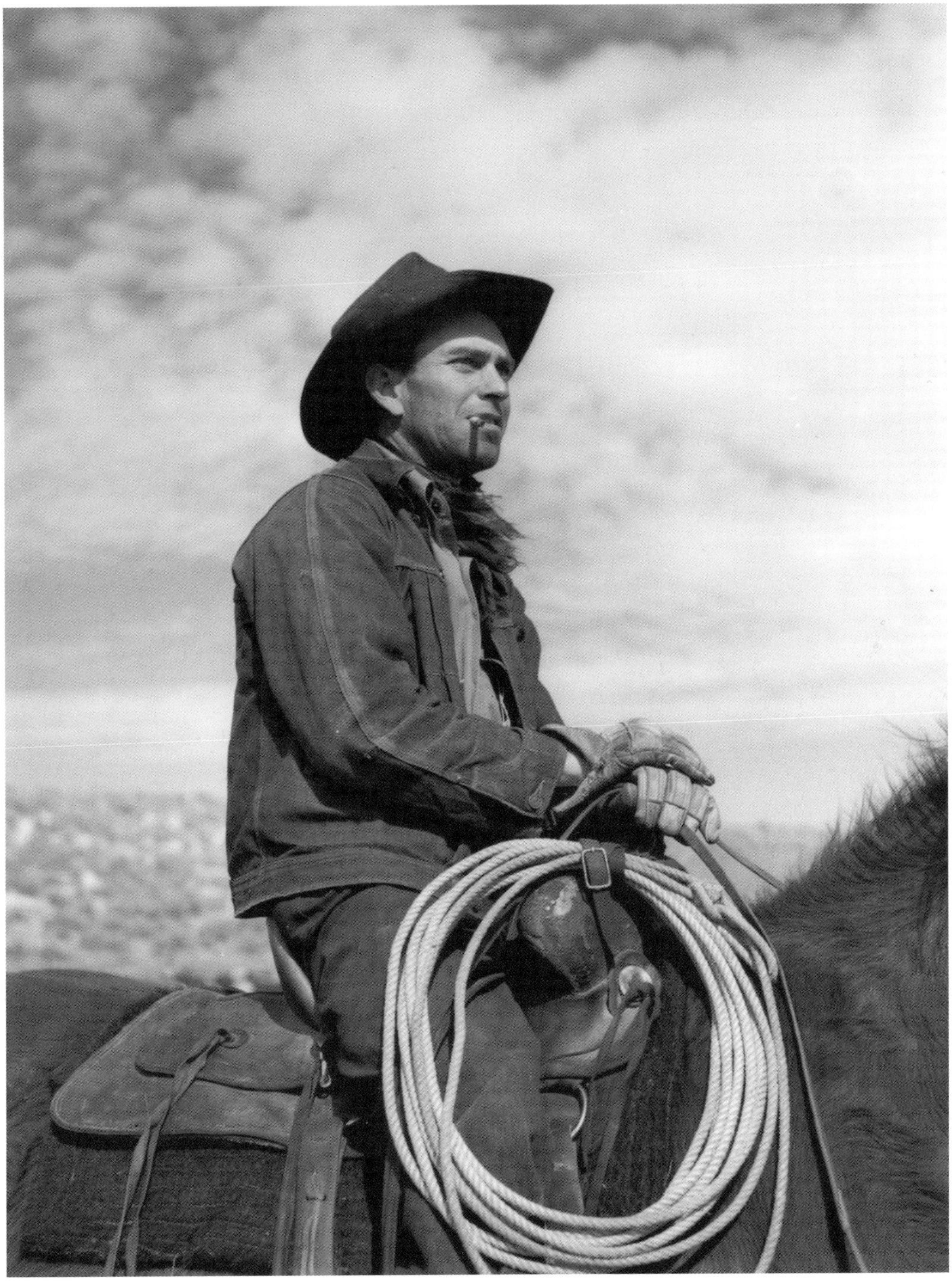

Cook at the Three-Circle Ranch roundup. Custer National Forest, Powder River County, Montana. Arthur Rothstein, June 1939.
LC-USF34-027535-D DLC

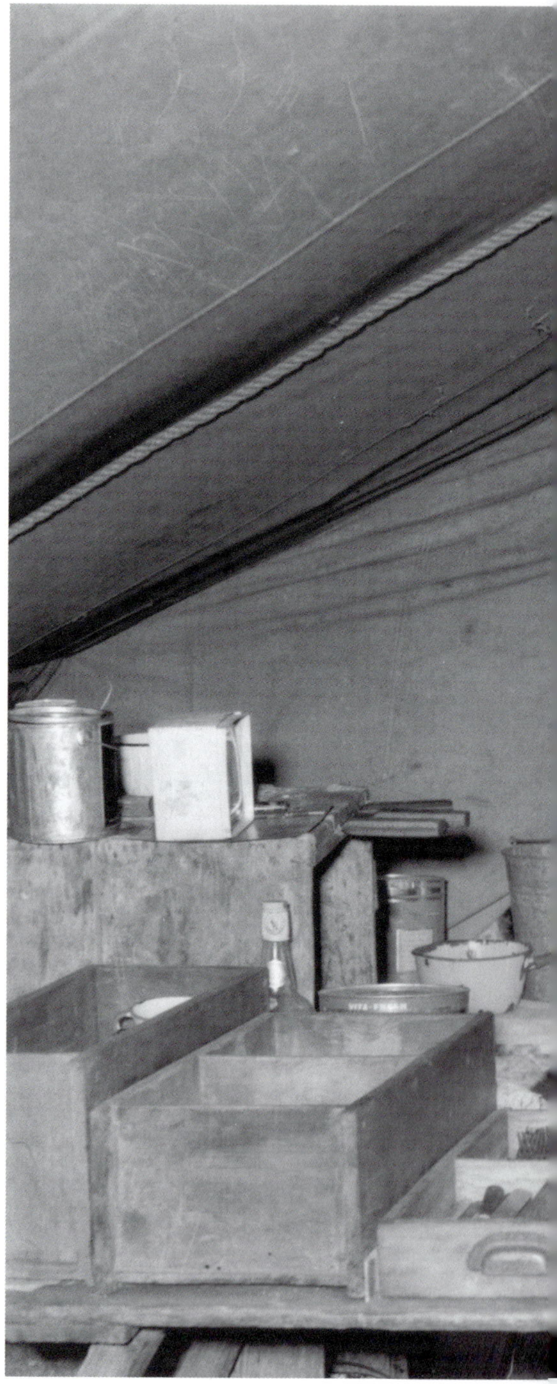

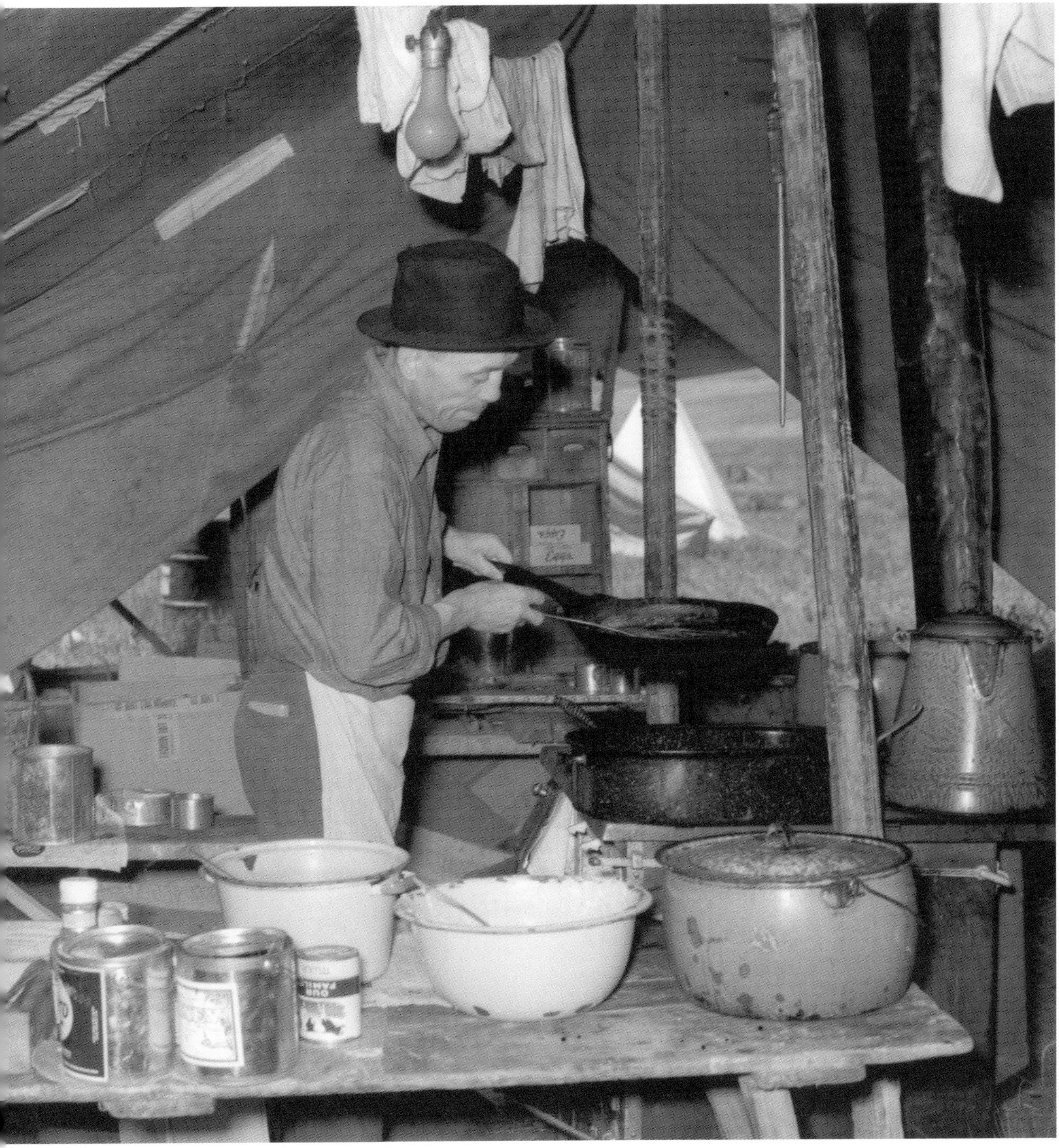

Cowhands at dinner. Quarter Circle U Ranch, Montana.
Arthur Rothstein, June 1939.
LC-USF34-027507-D

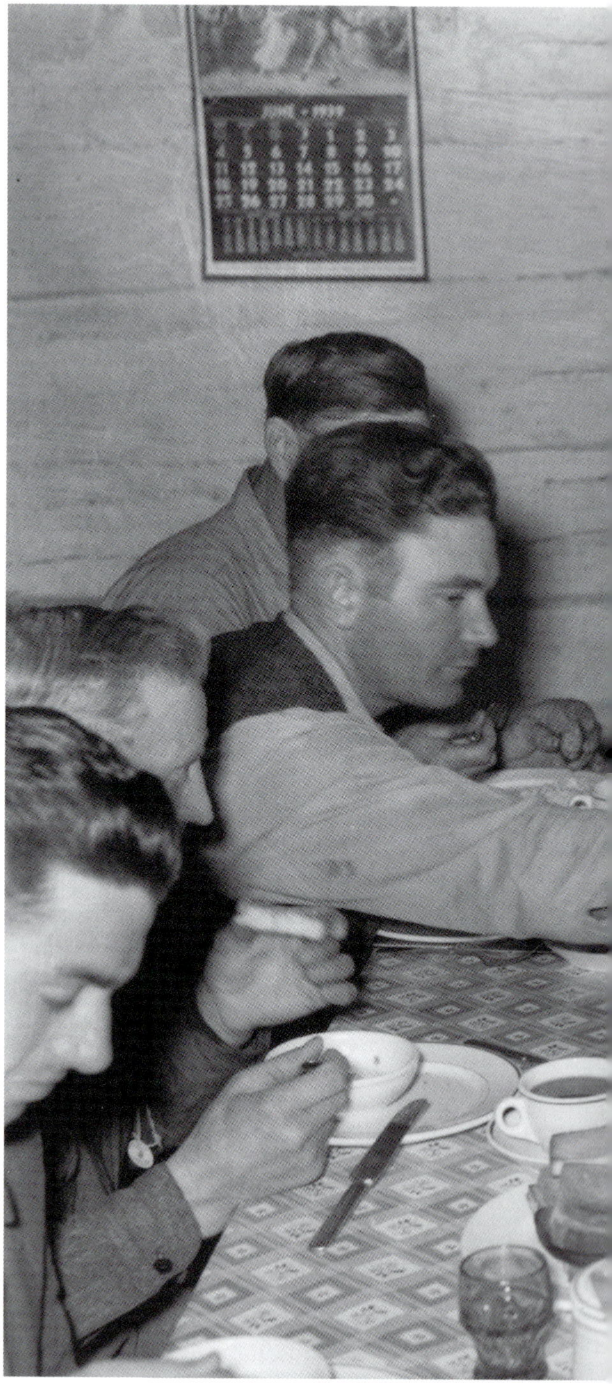

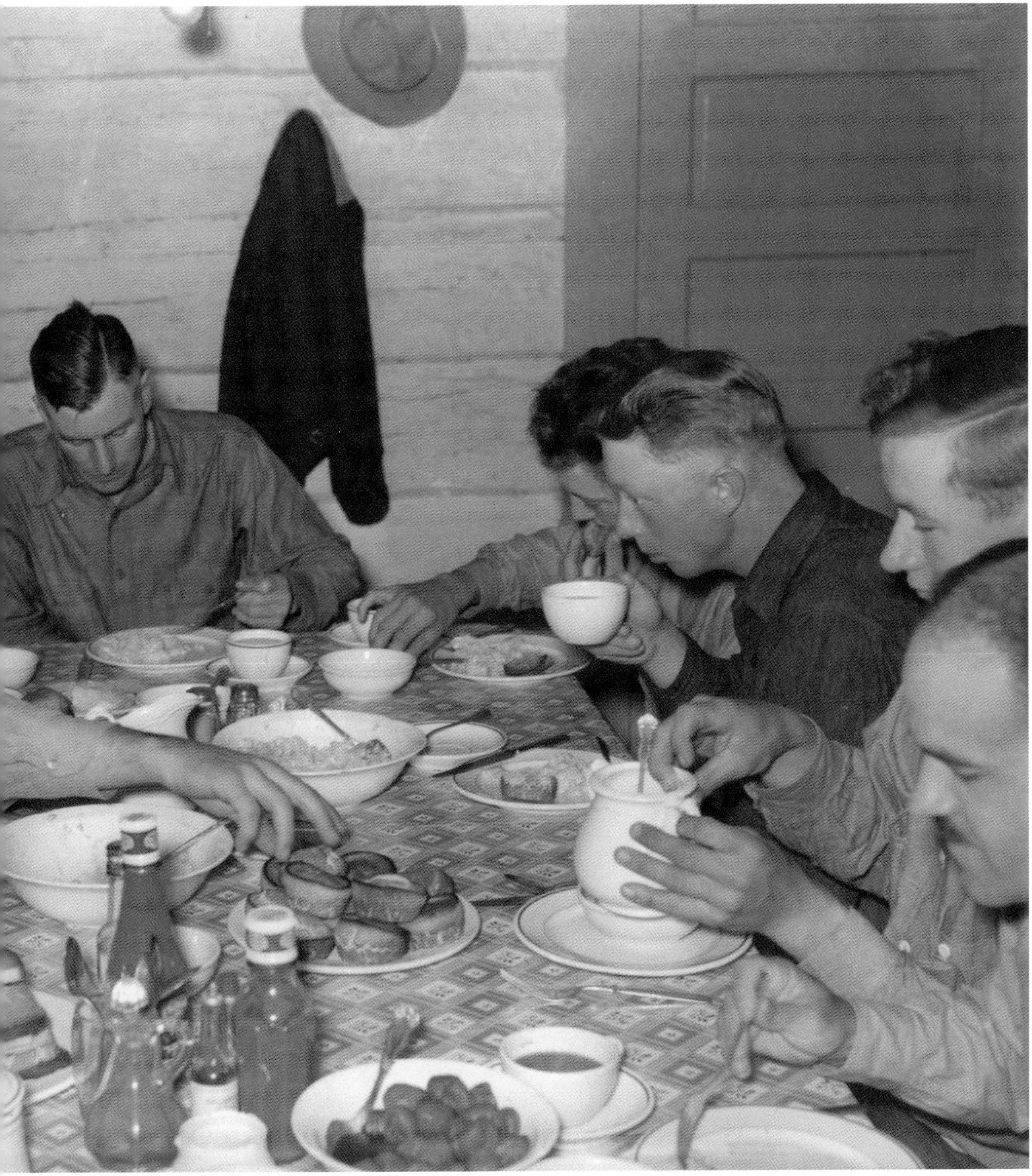

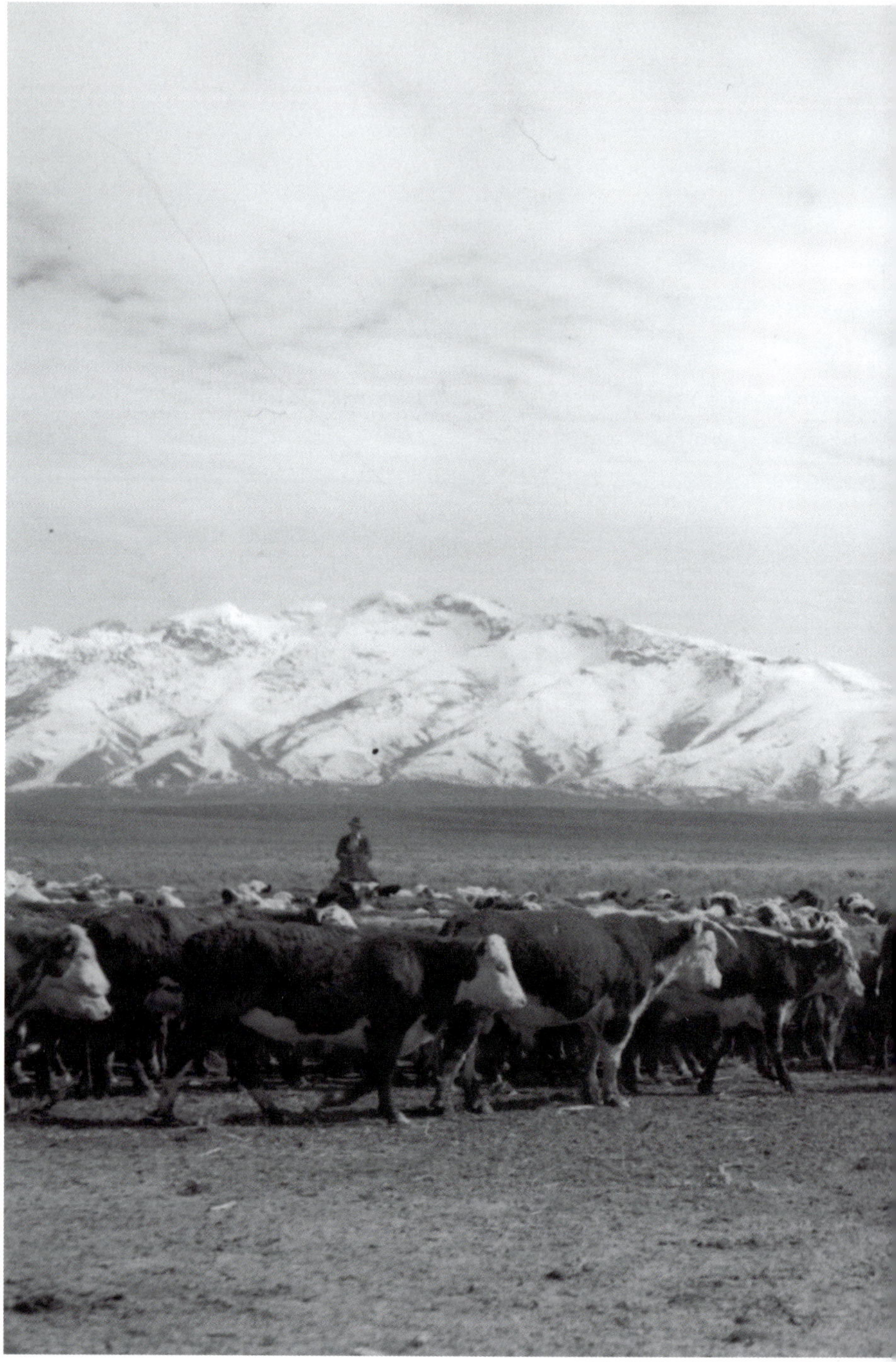

Rounding up of cattle. Elko County, Nevada. Arthur Rothstein, March 1940.
LC-USF34-029930-D DLC

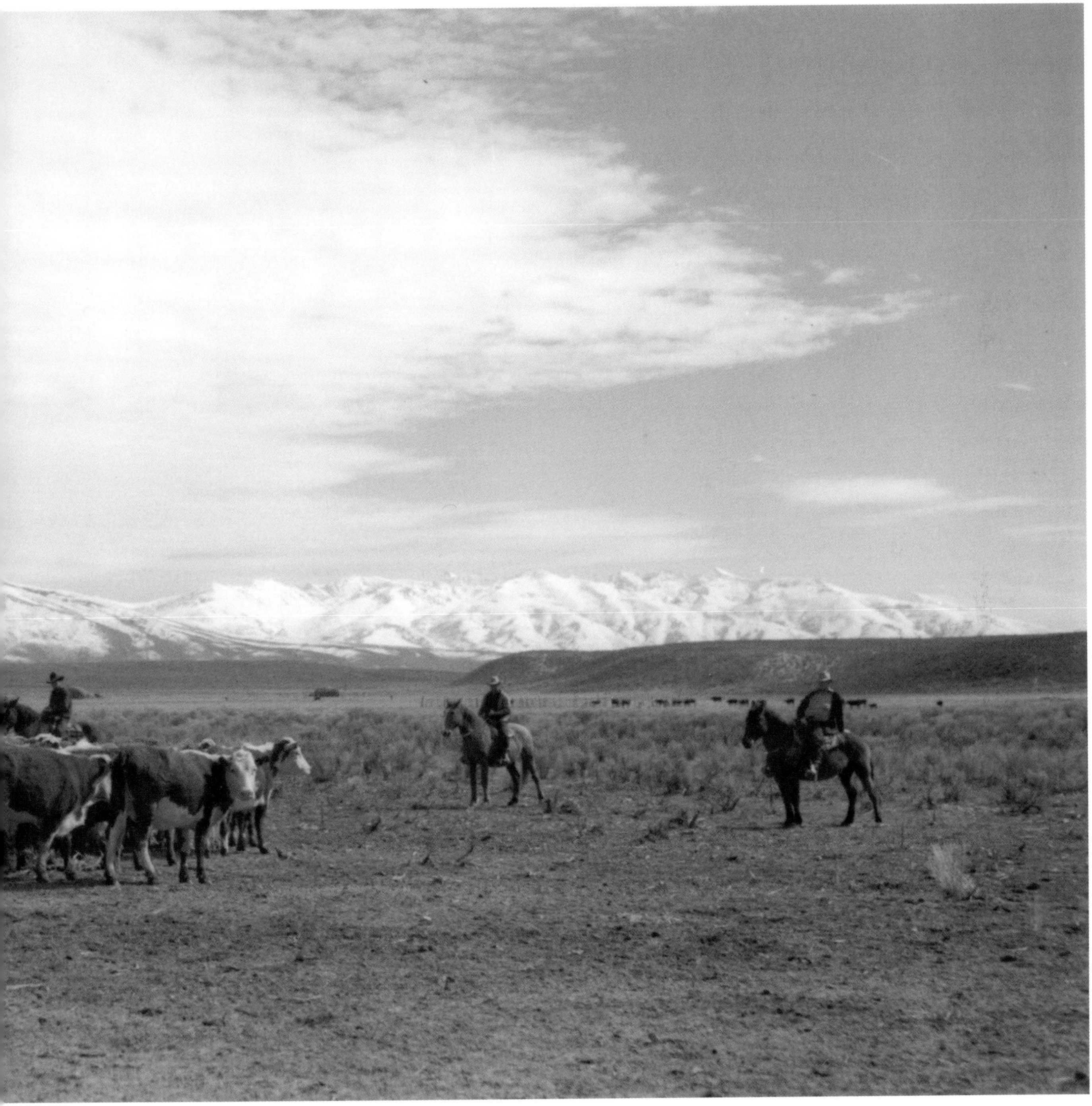

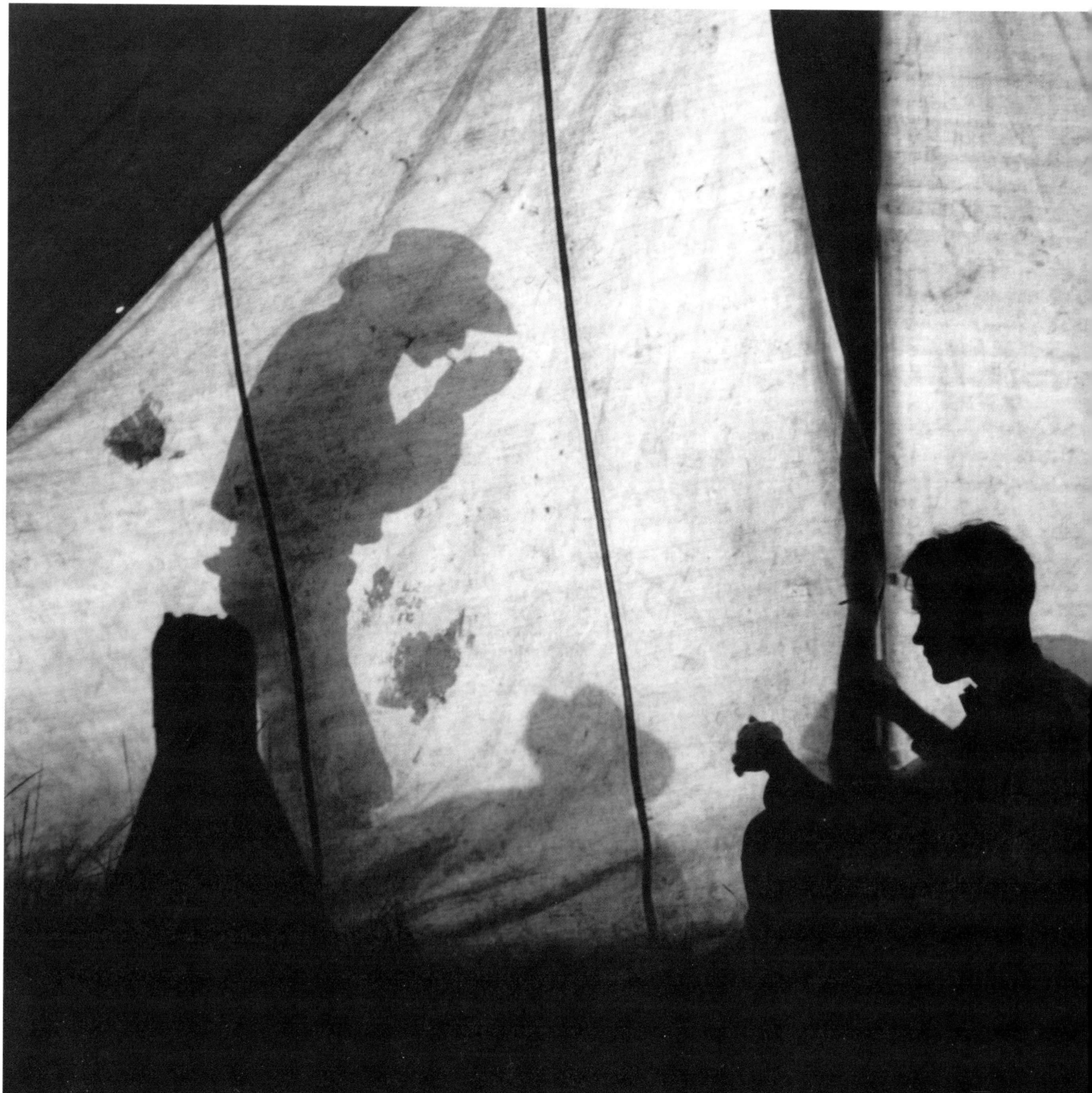

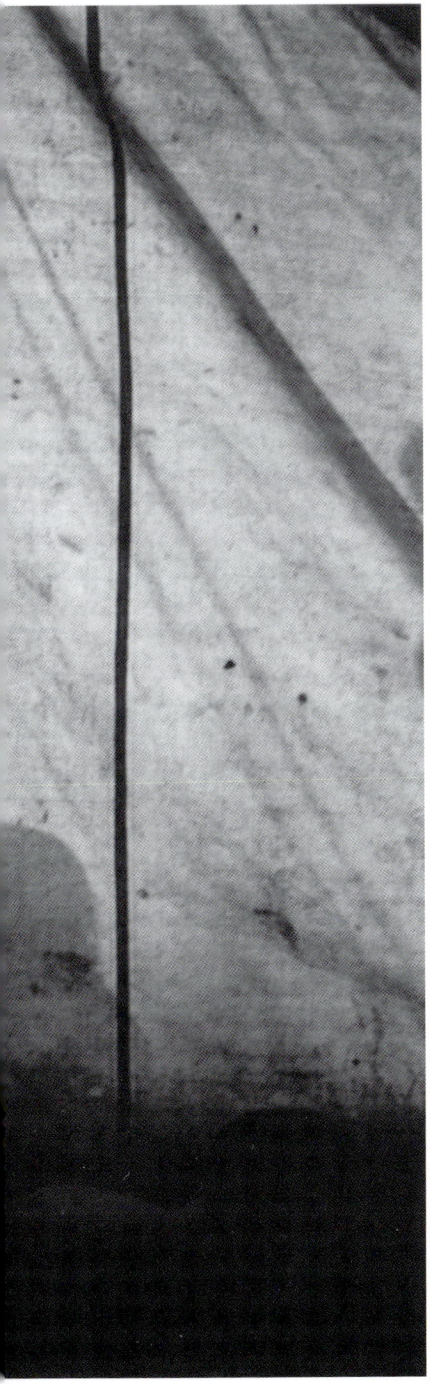

Shadows on a tent at the roundup. Quarter Circle U Ranch, Montana. Arthur Rothstein, June 1939.
LC-USF34-027679-D

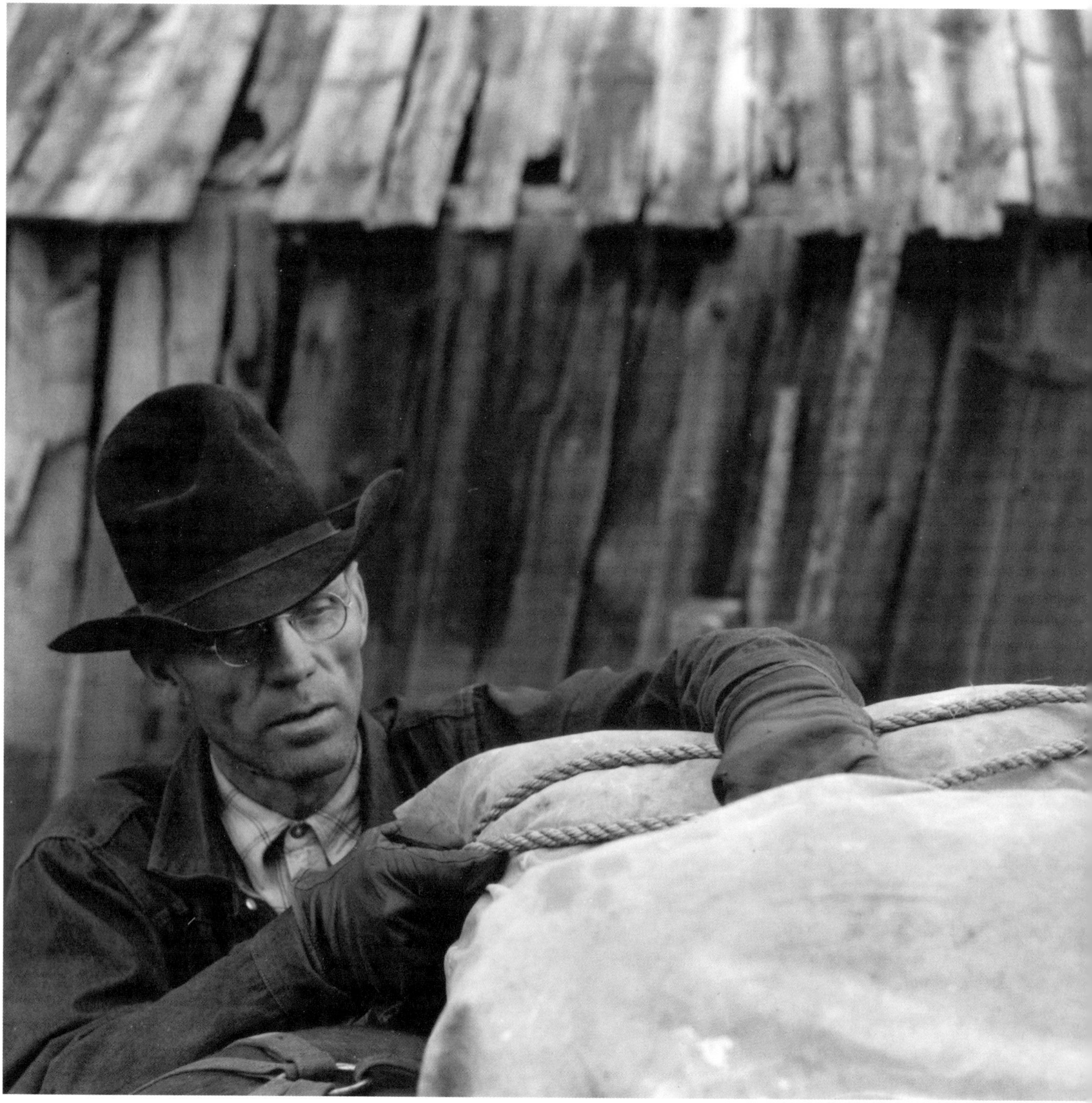

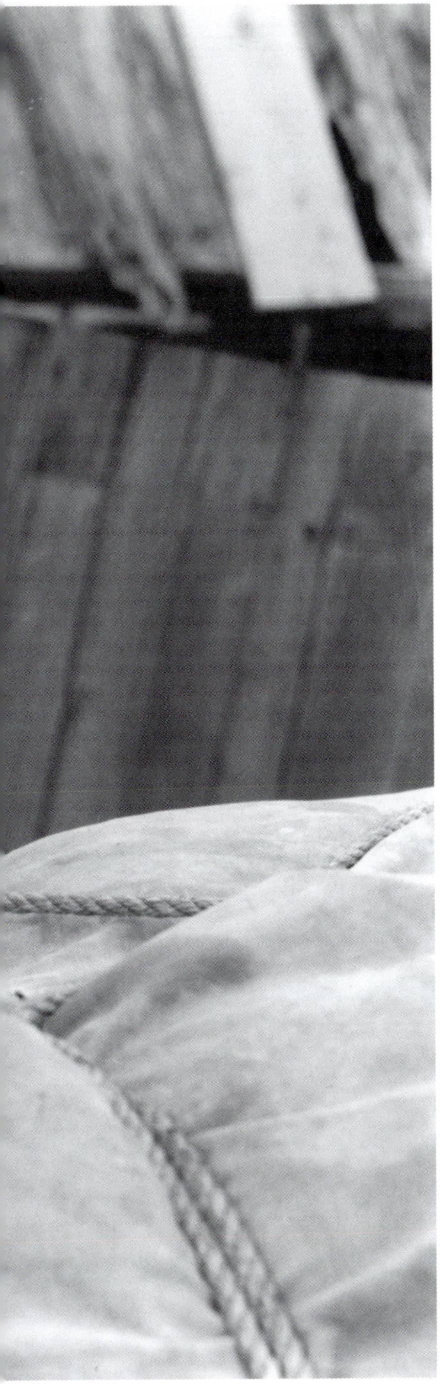

Using a diamond hitch in packing a horse. Bozeman, Montana. Arthur Rothstein, June 1939.
LC-USF34-027200-D DLC

Cowhand at a dance. Birney, Montana. Arthur Rothstein, June 1939.
LC-USF34-027496-D DLC; LC-USZ62-97083 DLC

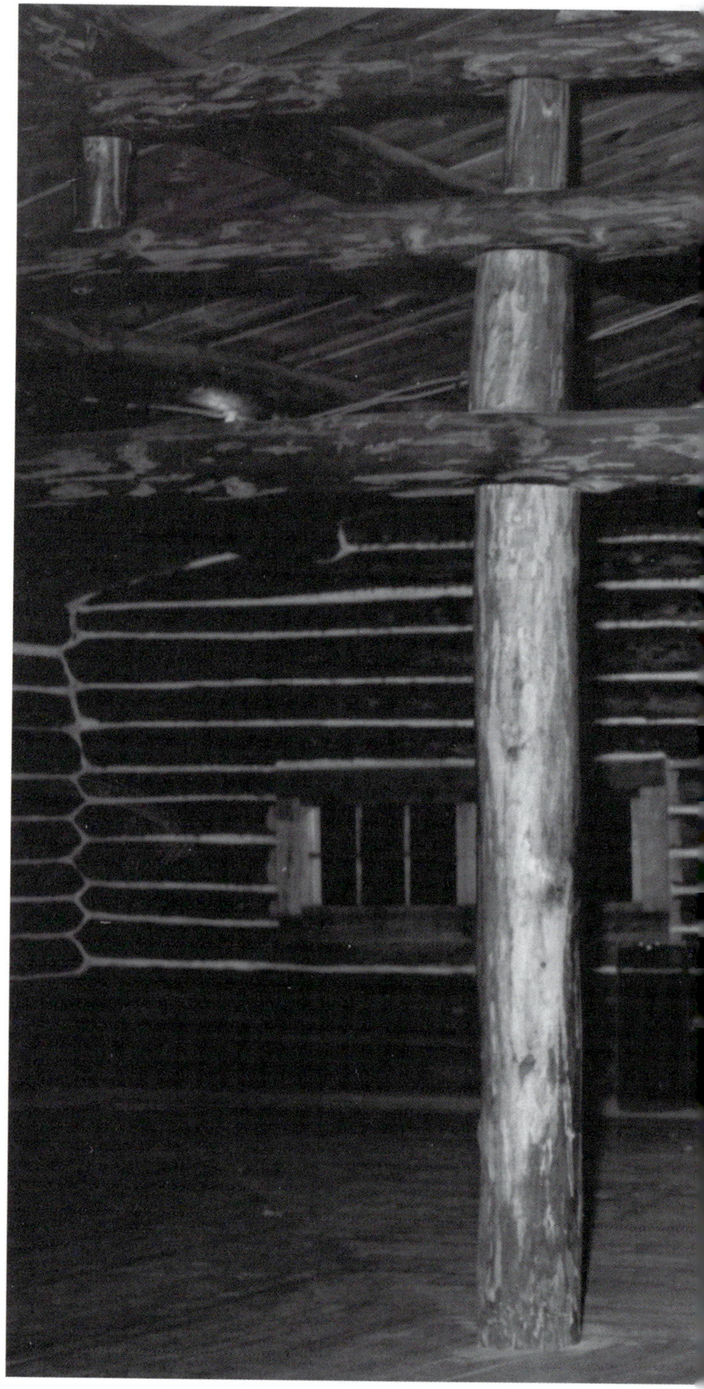

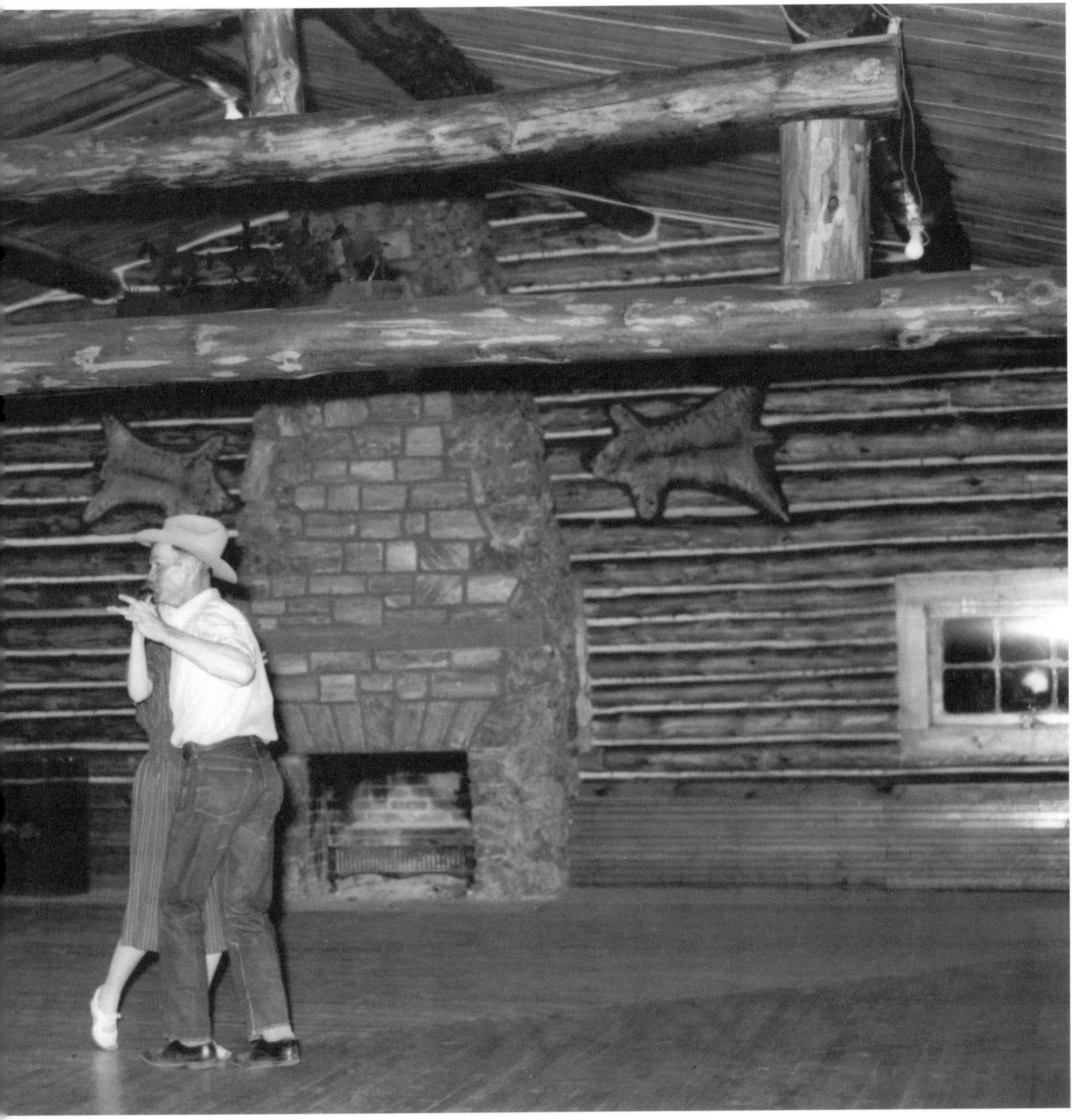

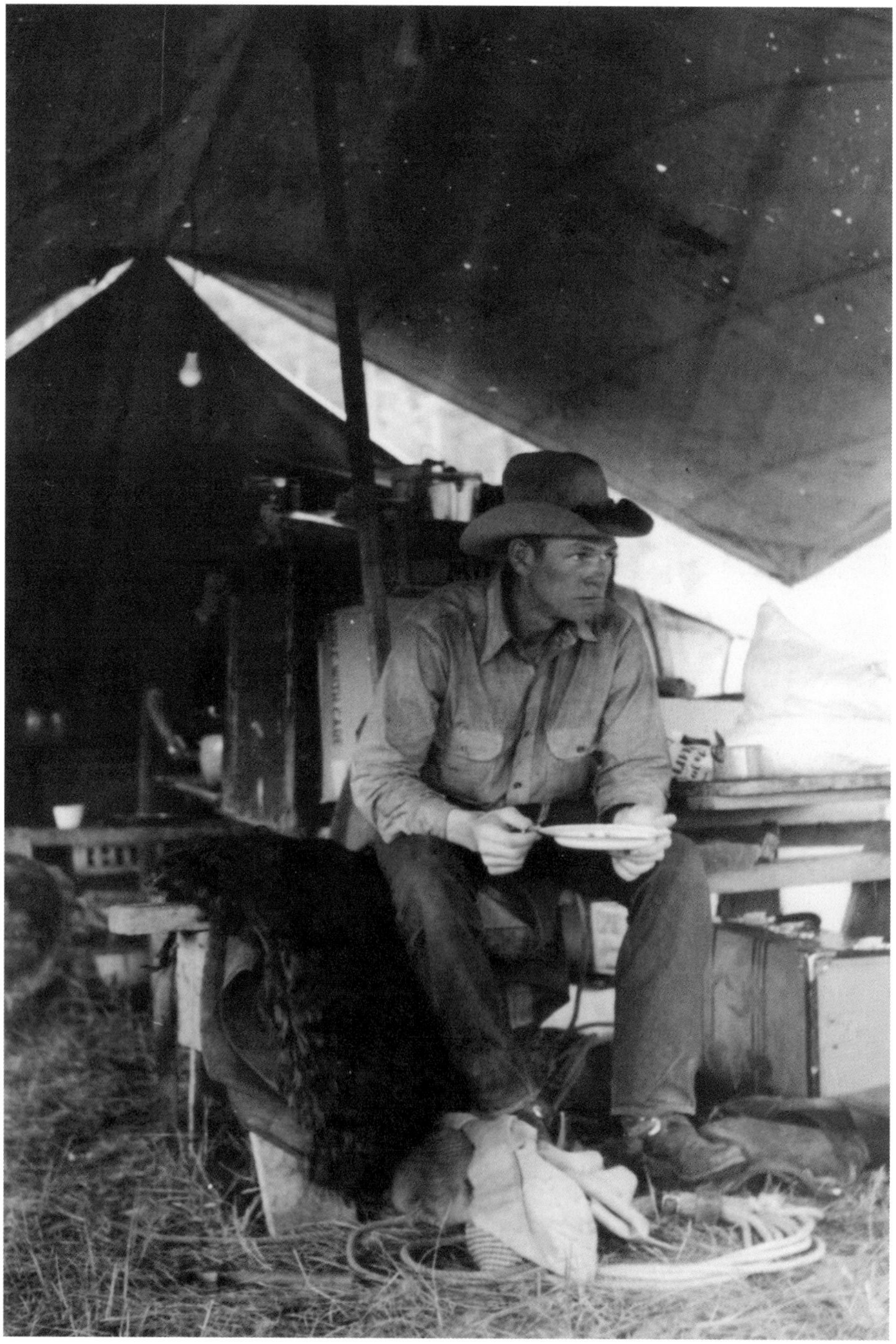

 Arthur Rothstein

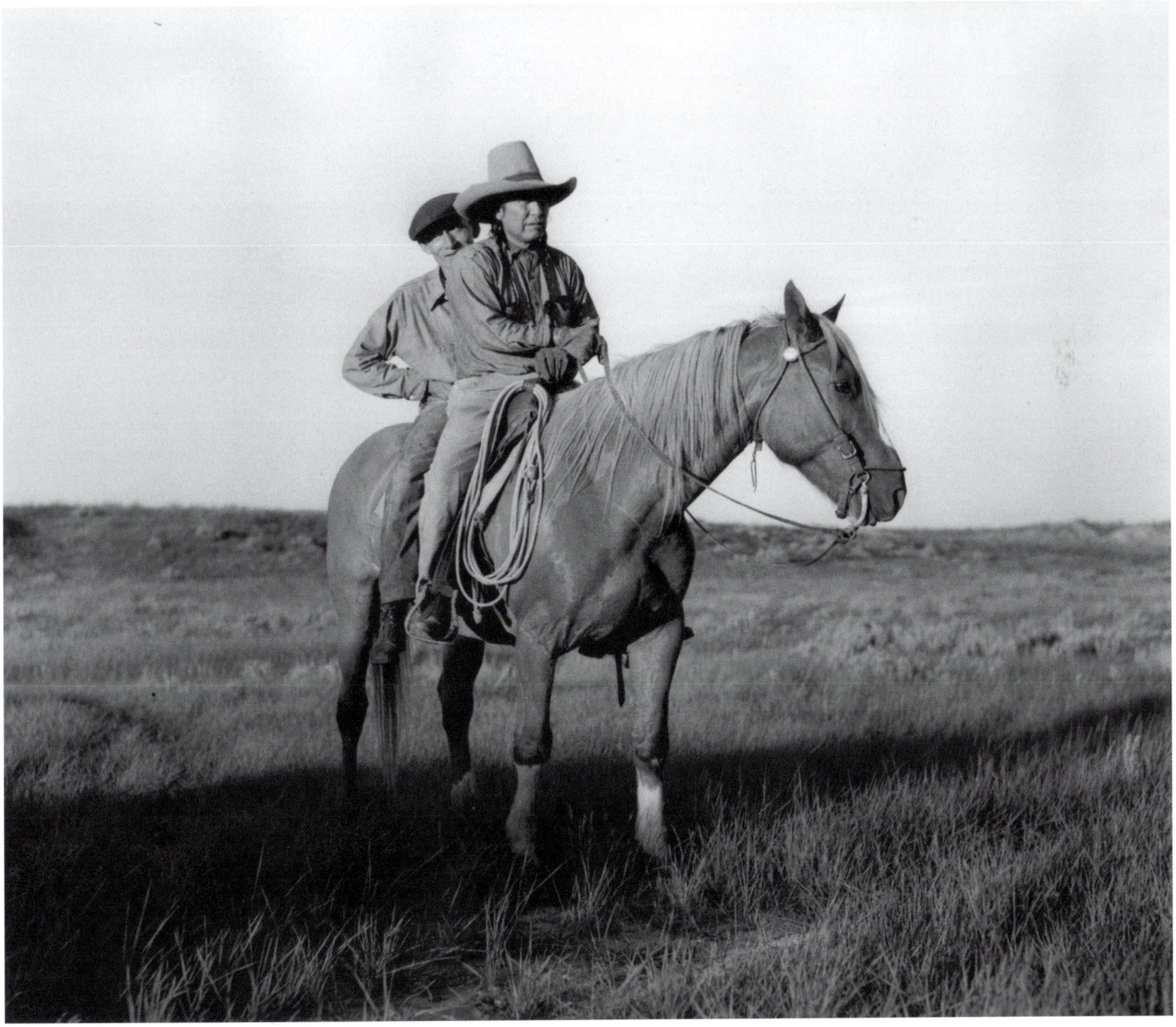

OPPOSITE: Cowboy eating in camp at the Three-Circle Ranch roundup. Custer National Forest, Powder River Country, Montana. Arthur Rothstein, June 1939.
LC-USF33-003166-M4 DLC

ABOVE: Cheyenne Indian and his son. Tongue River Reservation, Montana. Arthur Rothstein, June 1939.
LC-USF34-027684-D DLC

ELECTRICITY

The coming of electricity to ranches and farms in the rural West had a profound and far-reaching impact. The rural agricultural regions of the United States received electricity much later than the rest of the country. In 1910, about two percent of the farms and ranches in the United States had electricity. By 1920, only 600,000 of the country's 6.5 million farms and ranches were electrified. Because private electric companies didn't find it profitable to build lines and provide electricity to rural areas, many of the ranches in the West didn't finally get electricity until another of Franklin Roosevelt's New Deal programs, the Rural Electrification Administration, was established in 1935. By the end of the decade, forty percent of rural homes had electricity.

The lives of ranching families were transformed by access to electricity. Working time was extended beyond dawn and dusk, adding two to four hours to the day. No longer did a rancher have to carry a lantern into the barn, risking fire and limiting him to working with only his other hand. With electricity came appliances, such as refrigerators, allowing preservation of foods, and electric cook stoves, eliminating the need to collect firewood for cooking. Electric pumps provided water for indoor plumbing, and water no longer had to be hauled from the manual pump outside to the house in numerous trips each day. Electric pumps also made the watering of stock much more efficient. A horse or cow can drink as much as twenty-five gallons a day, and a pump could deliver one-thousand gallons a day for about five cents. Electricity increased productivity, decreased manual labor, and improved sanitation and health.

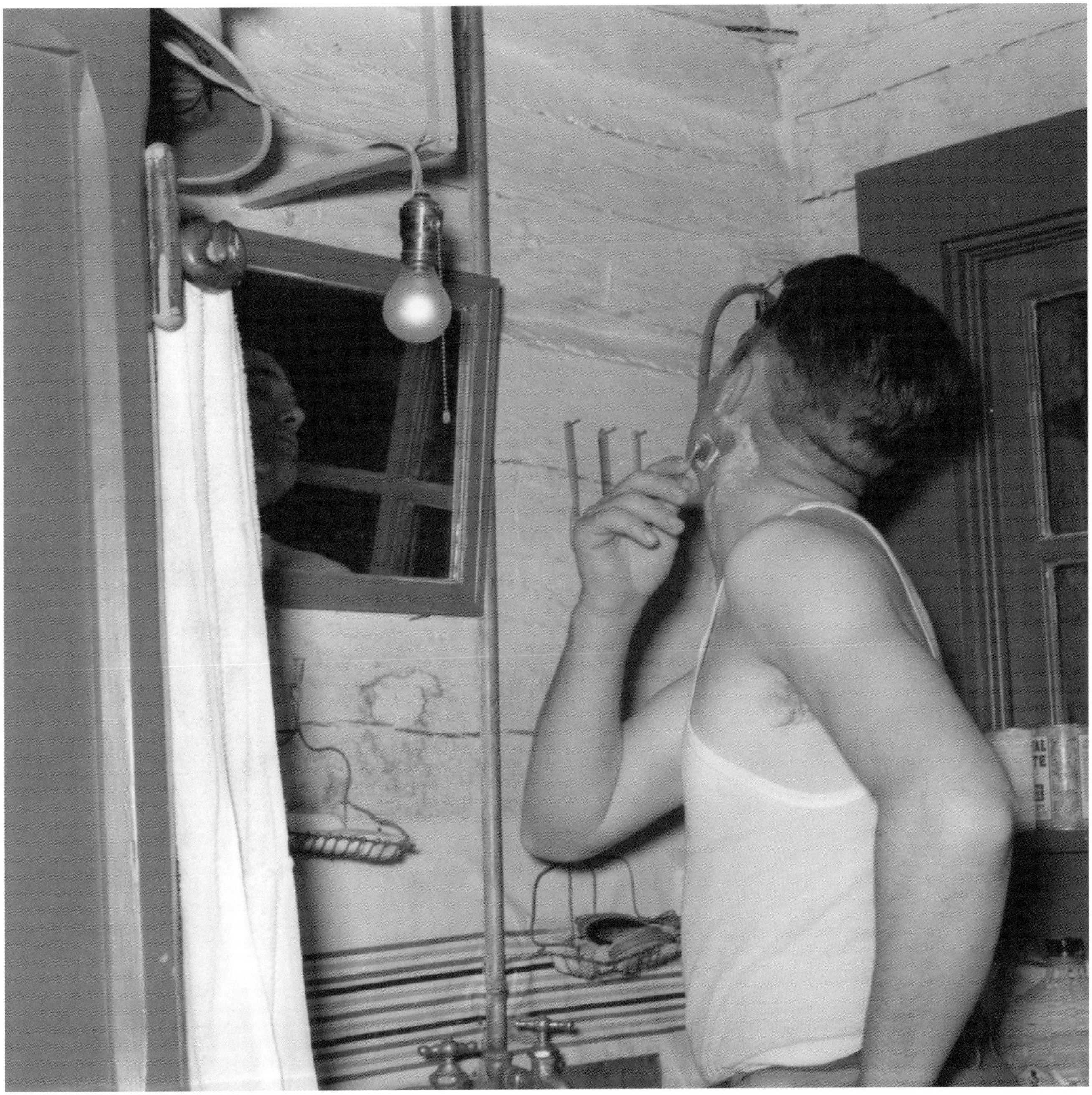

Cowhand shaving. Quarter Circle U Ranch, Montana. Arthur Rothstein, June 1939.
LC-USF34-027572-D DLC

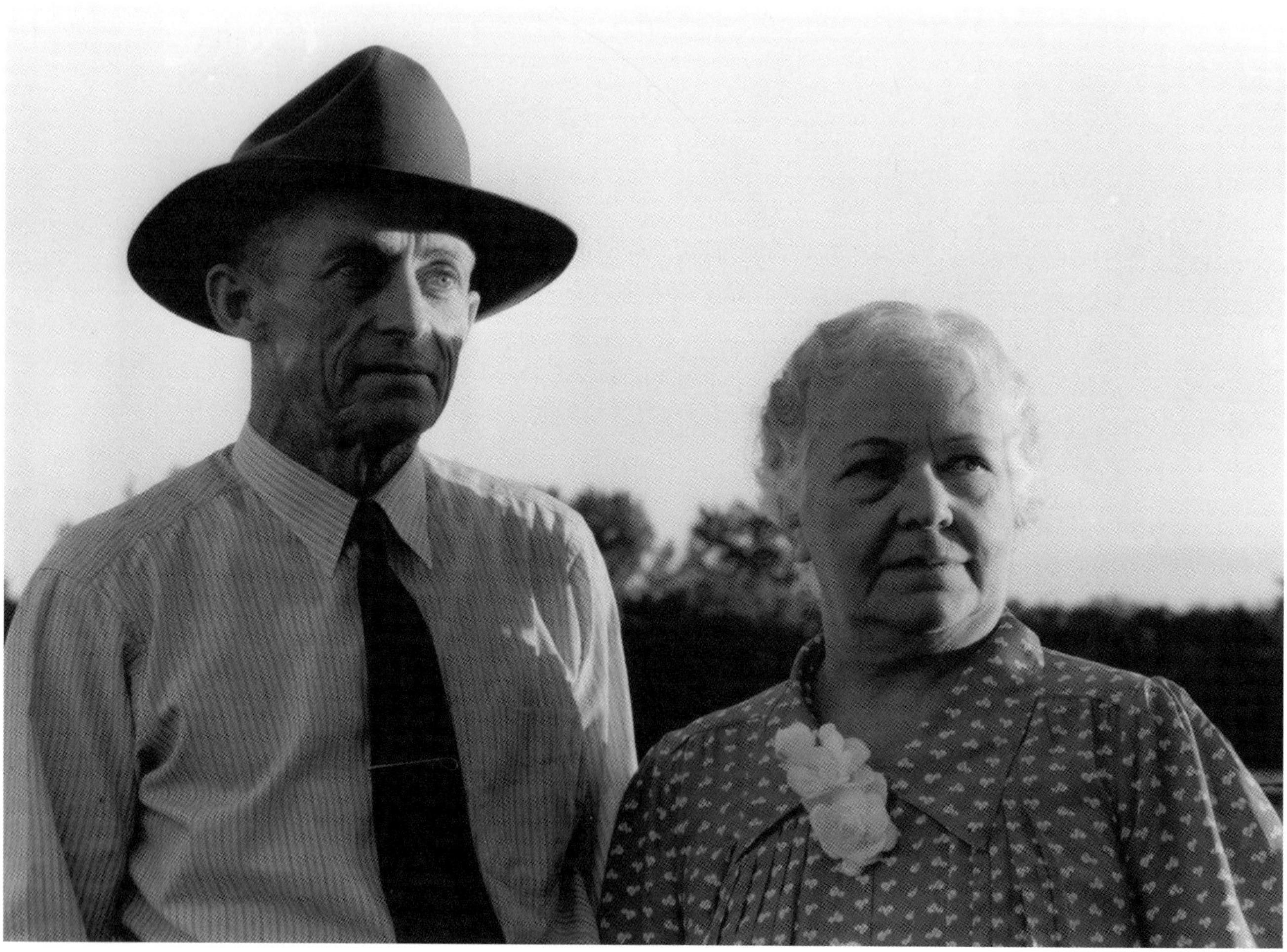

Mr. and Mrs. Jack Arnold, owners. Quarter Circle U Ranch, Montana. Arthur Rothstein, June 1939.
LC-USF34-027587-D DLC

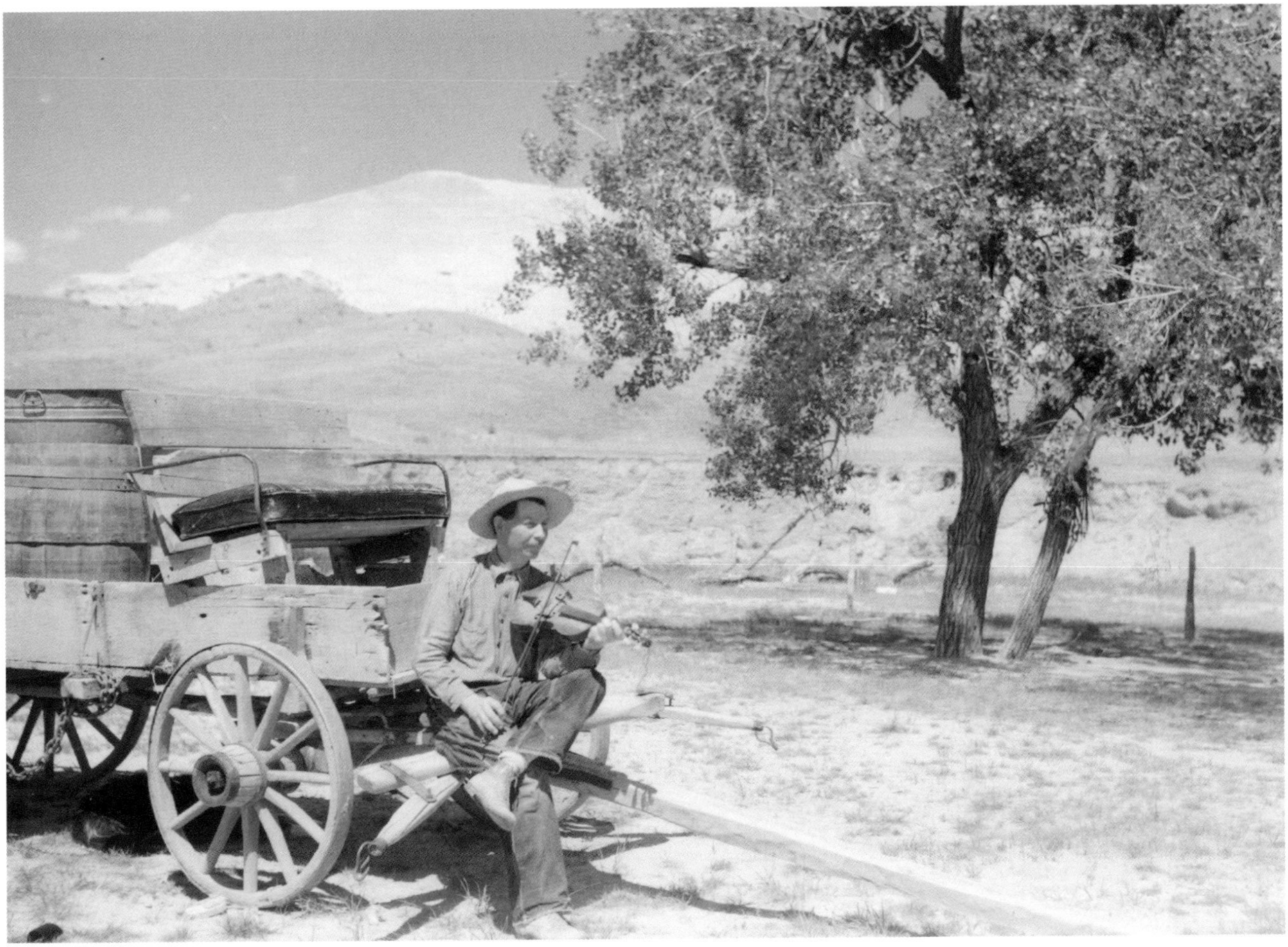

Homesteader in the land development area. Pennington County, South Dakota. Arthur Rothstein, May 1936.
LC-USF34-004512-D DLC

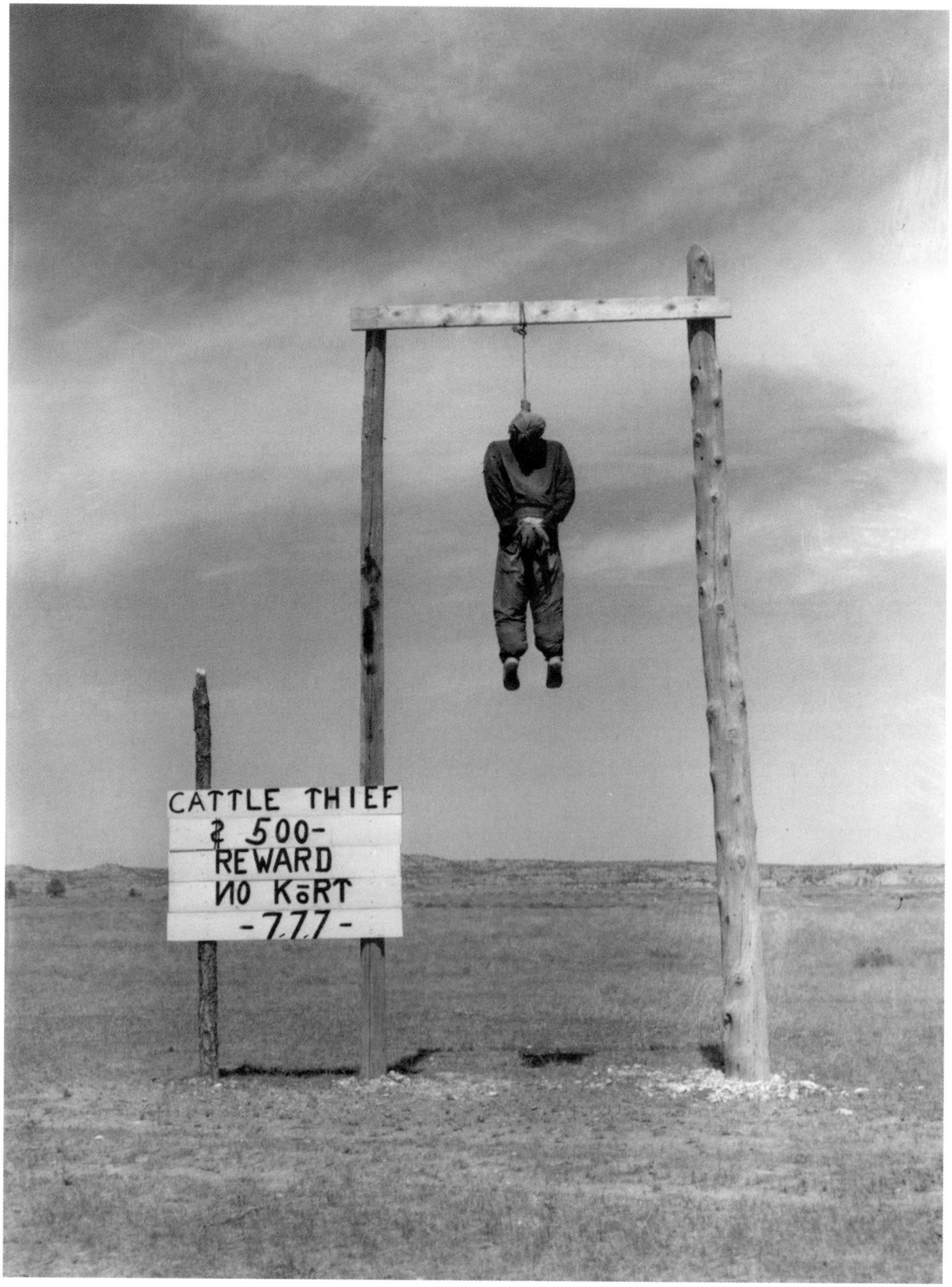
CATTLE THIEF
$ 500-
REWARD
NO KoRT
-777-

OPPOSITE: Cattle thief hung in effigy along US Highway 10 to provide Western atmosphere for tourists. "777" refers to a secret password of vigilantes in 1864. Custer County, Montana. Arthur Rothstein, June-July 1939. *LC-USF34-027239-D DLC*

RIGHT: Cowboy with amateur movie camera. Quarter Circle U Ranch roundup. Montana. Arthur Rothstein, June 1939. *LC-USF33-003217-M1 DLC*

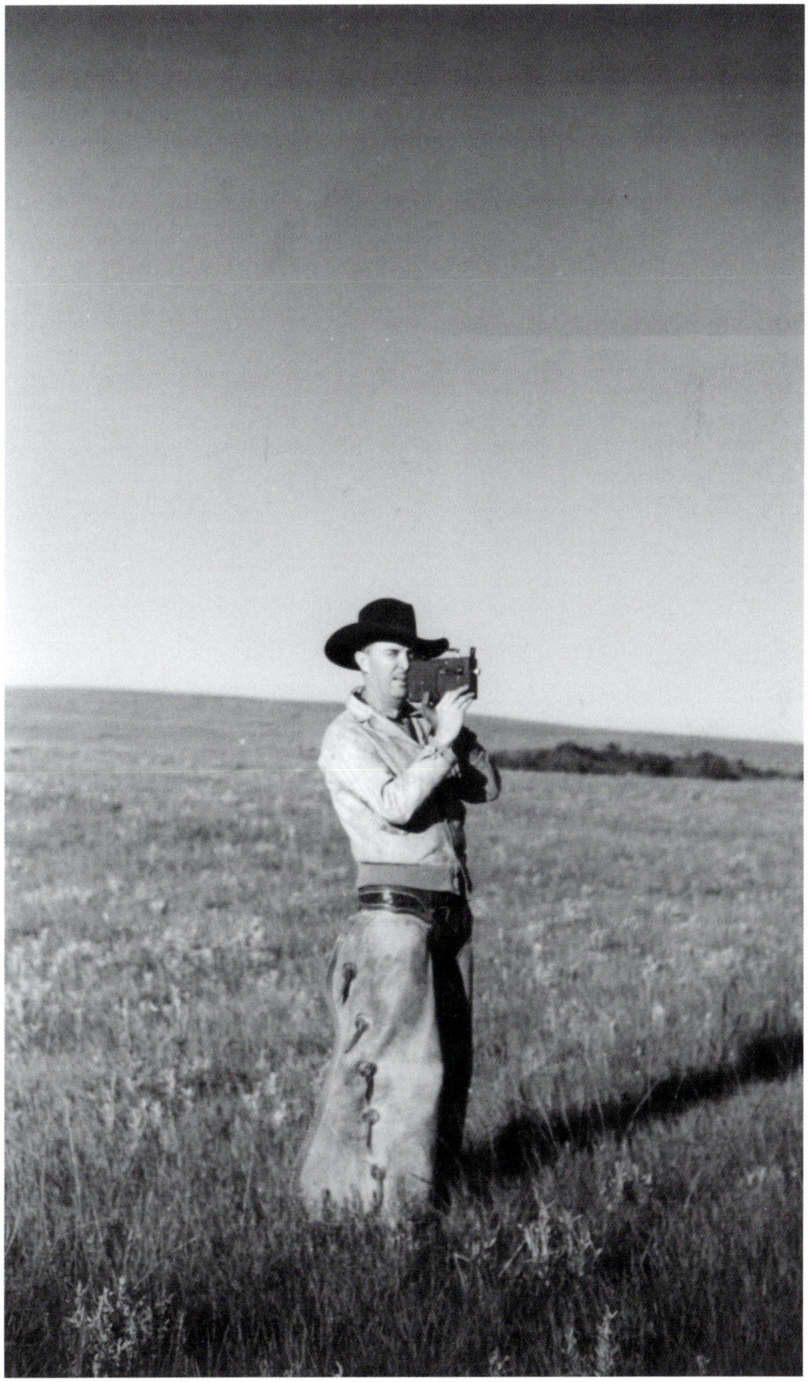

Beer parlor. Birney, Montana. Arthur Rothstein, June 1939.
LC-USF33-003160-M3 DLC

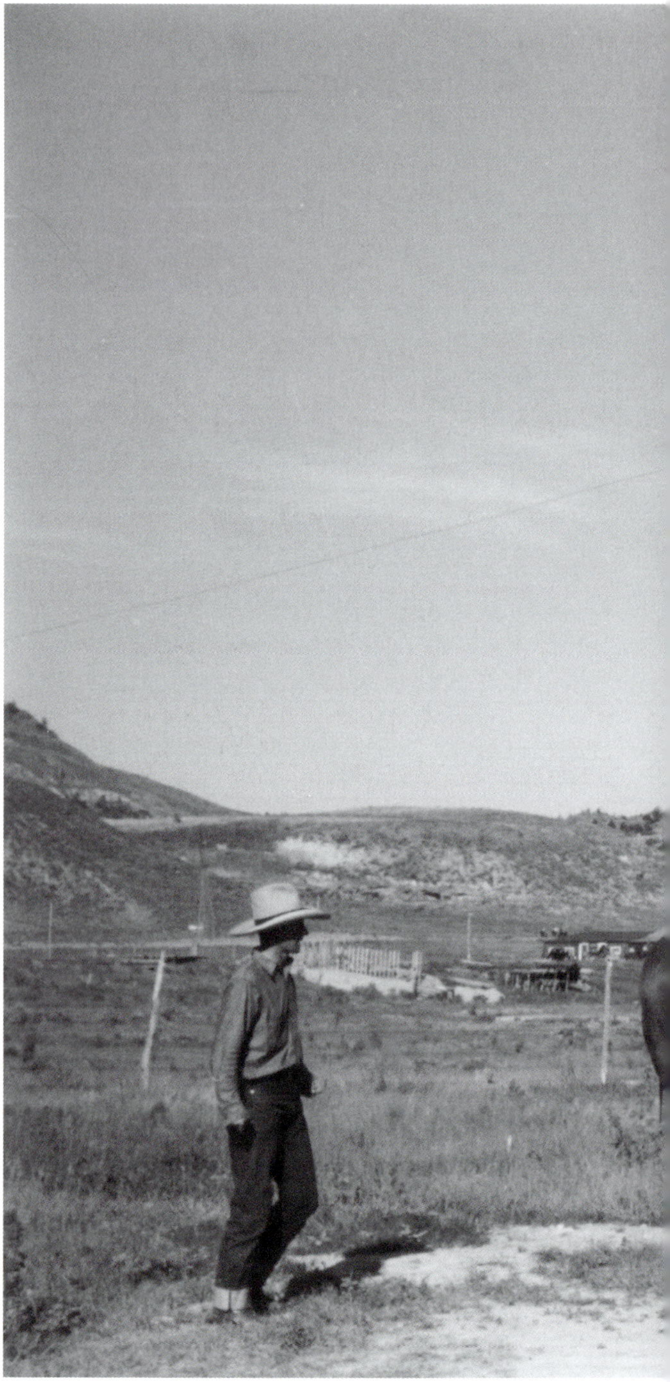

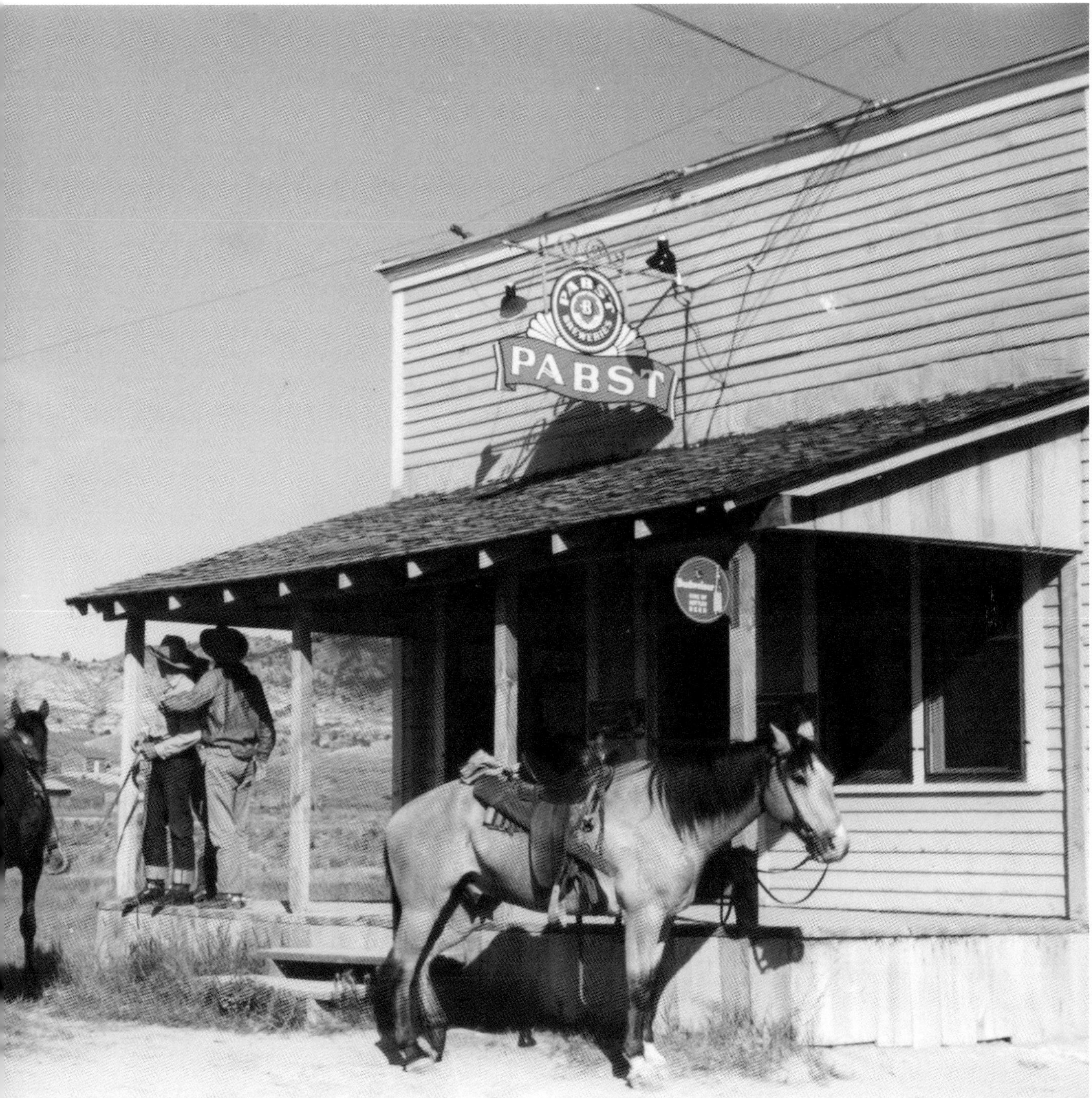
PABST
BREWERIES
PABST

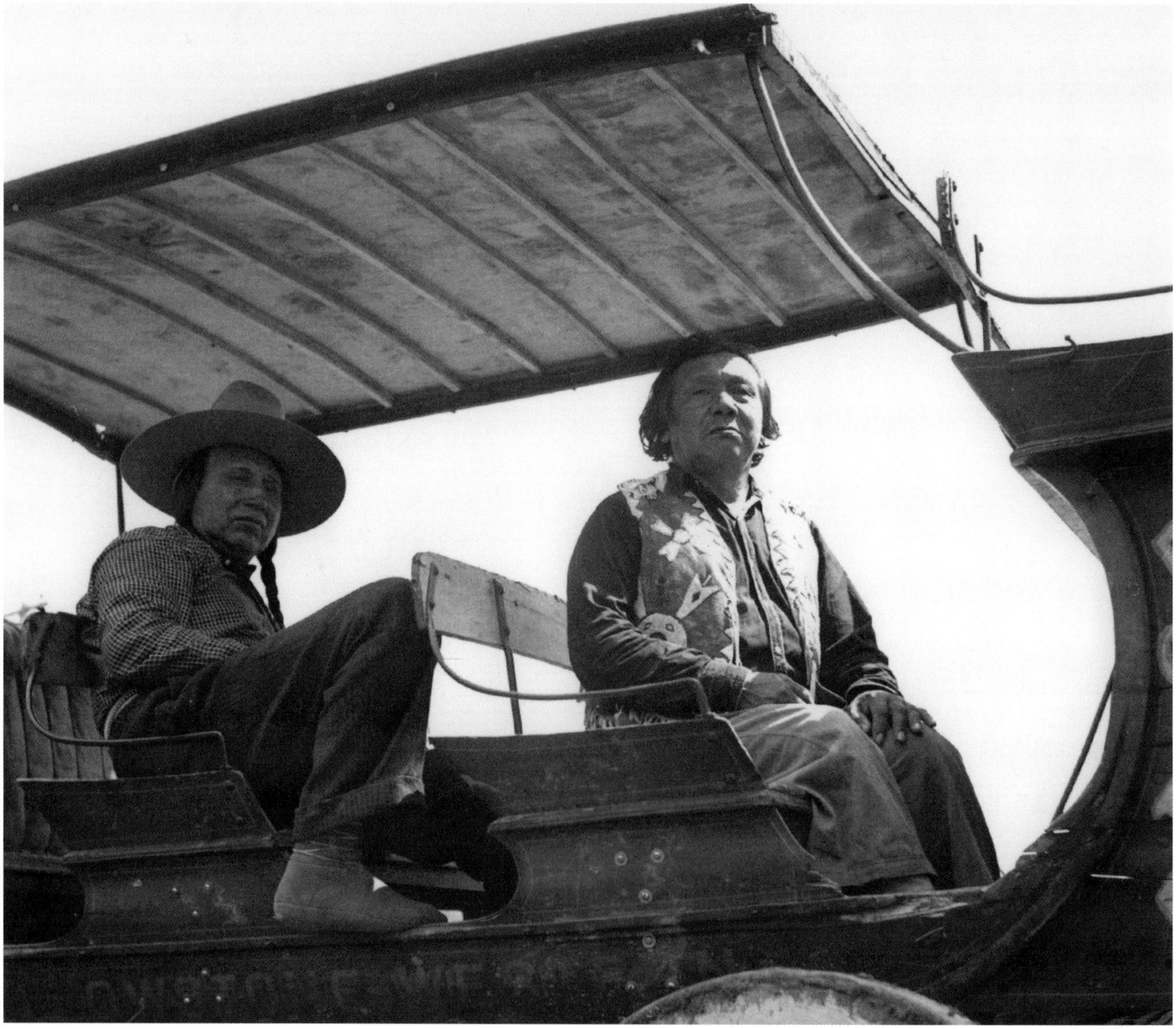

ABOVE: Indians at the Mollala Buckaroo (rodeo). Arthur Rothstein, July 1936.
LC-USF34-004965-E DLC

OPPOSITE: Cowhand. Quarter Circle U Brewster-Arnold ranch company, near Birney, Montana. Arthur Rothstein, June 1939.
LC-USF34-027550-D DLC.

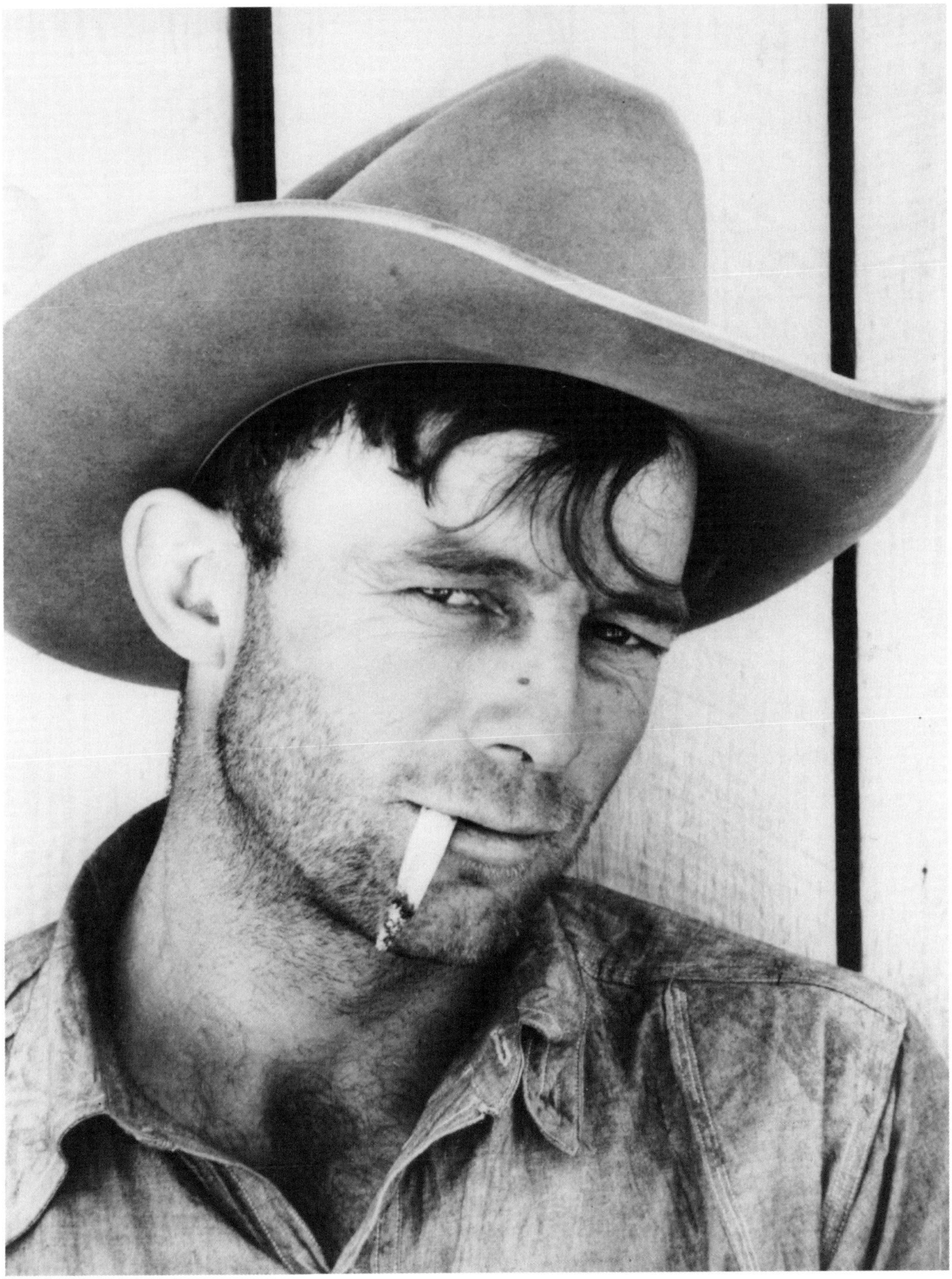

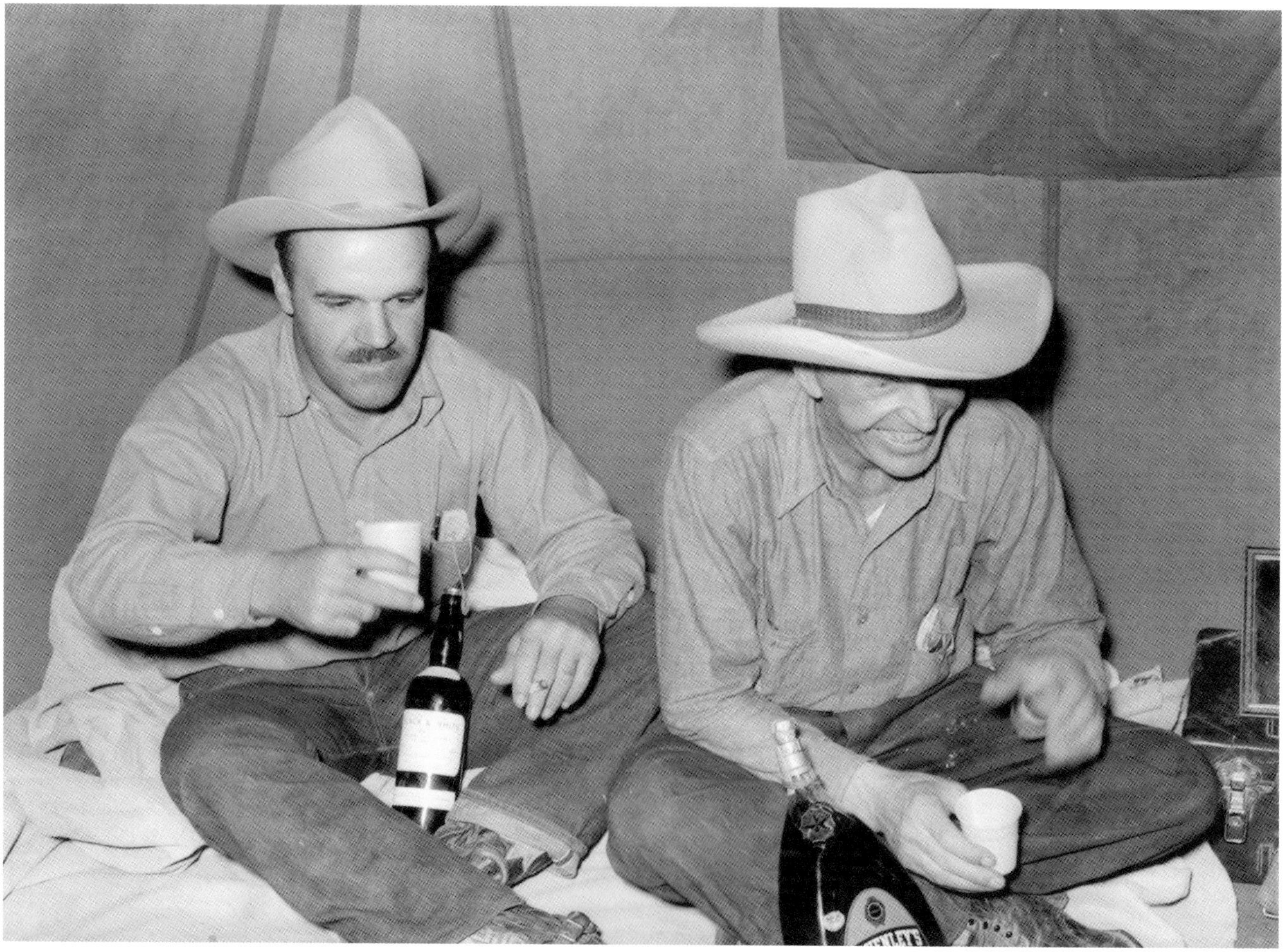

Cowhands relaxing at the Quarter Circle U Ranch roundup. Big Horn County, Montana. Arthur Rothstein, June 1939.
LC-USF34-027724-D DLC

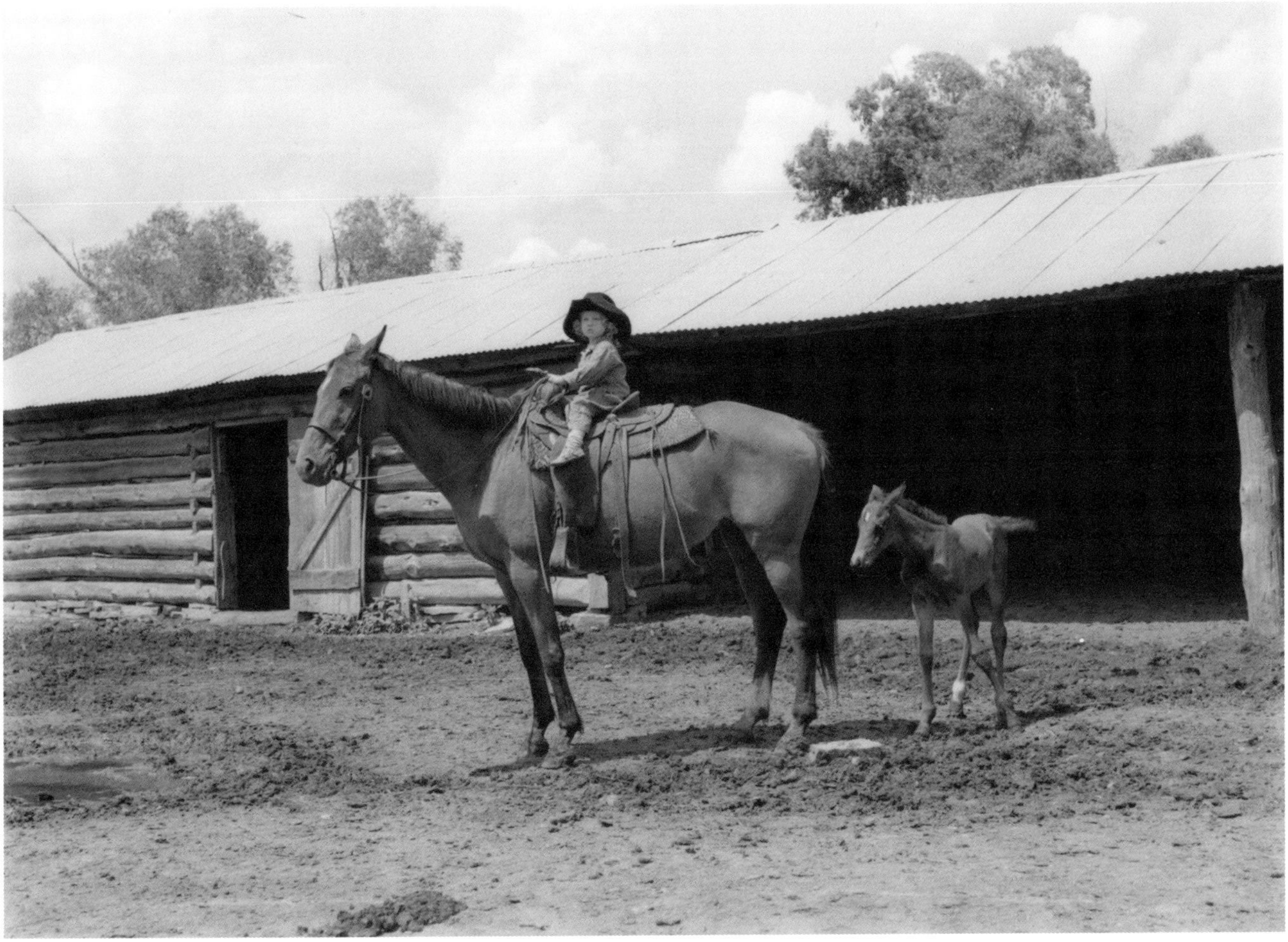

Little girl with mare. Warren Brewster ranch, Montana. Arthur Rothstein, June 1939.
LC-USF34-027659-D DLC

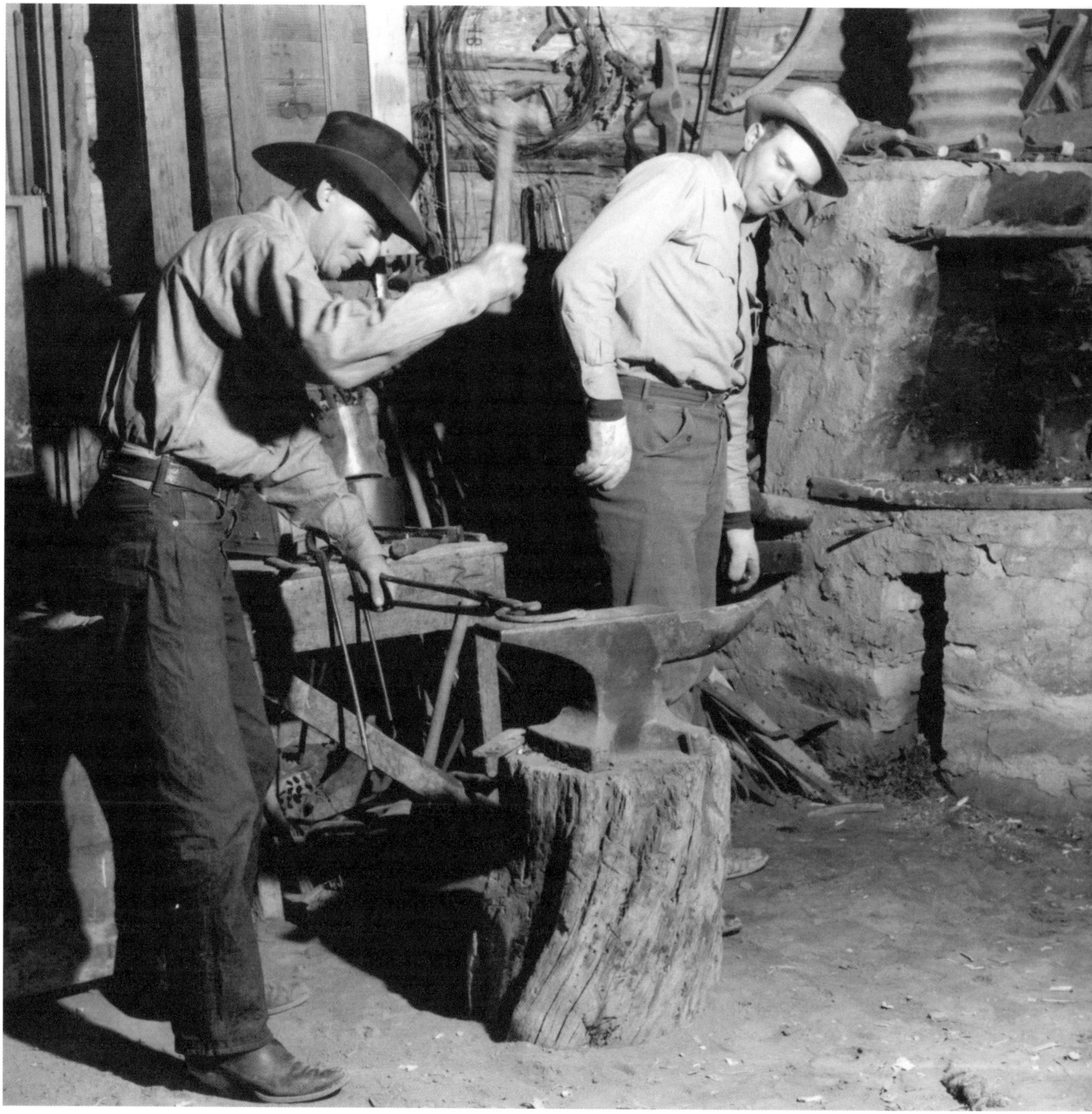

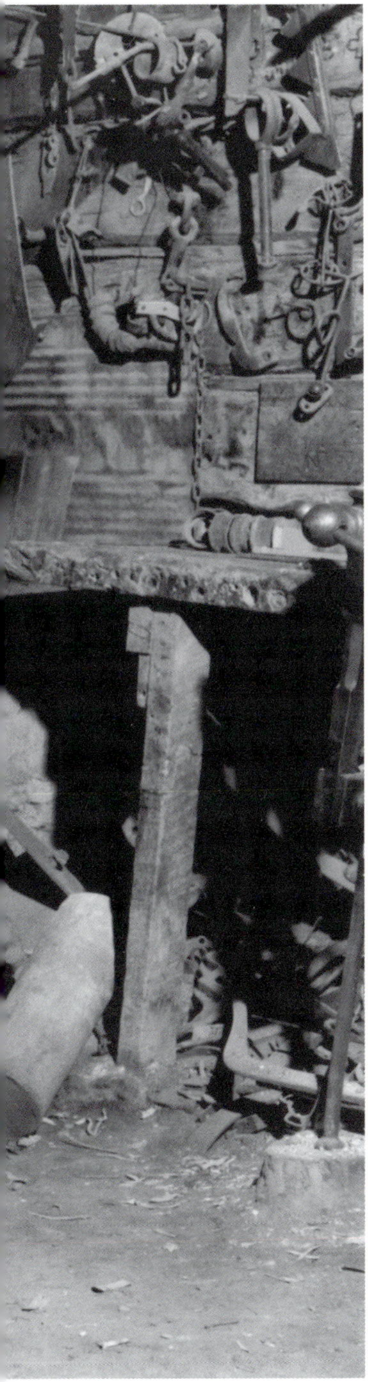

Blacksmith shop. Quarter Circle U Ranch, Montana. Arthur Rothstein, June 1939.
LC-USF34-027638-D DLC

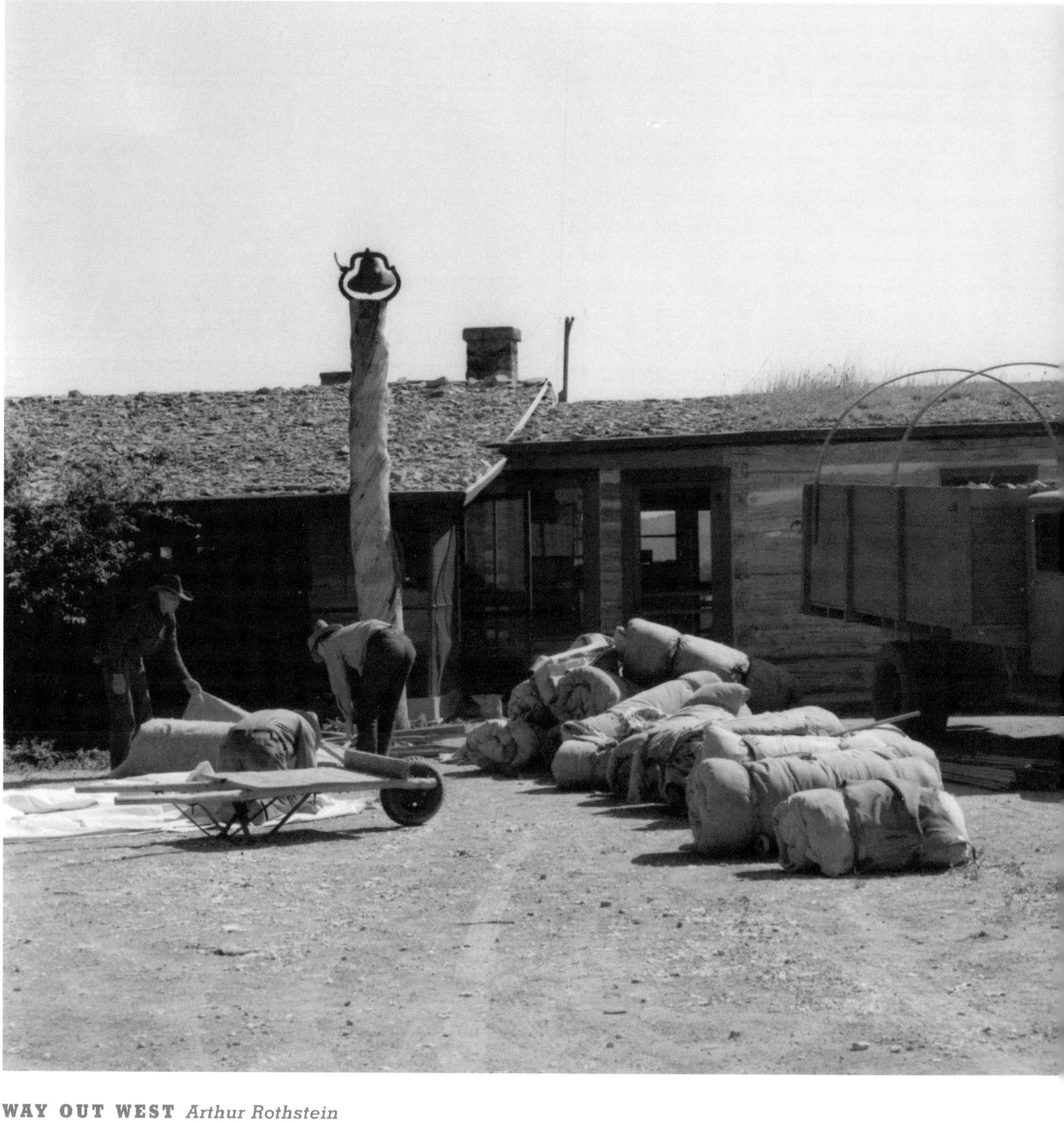

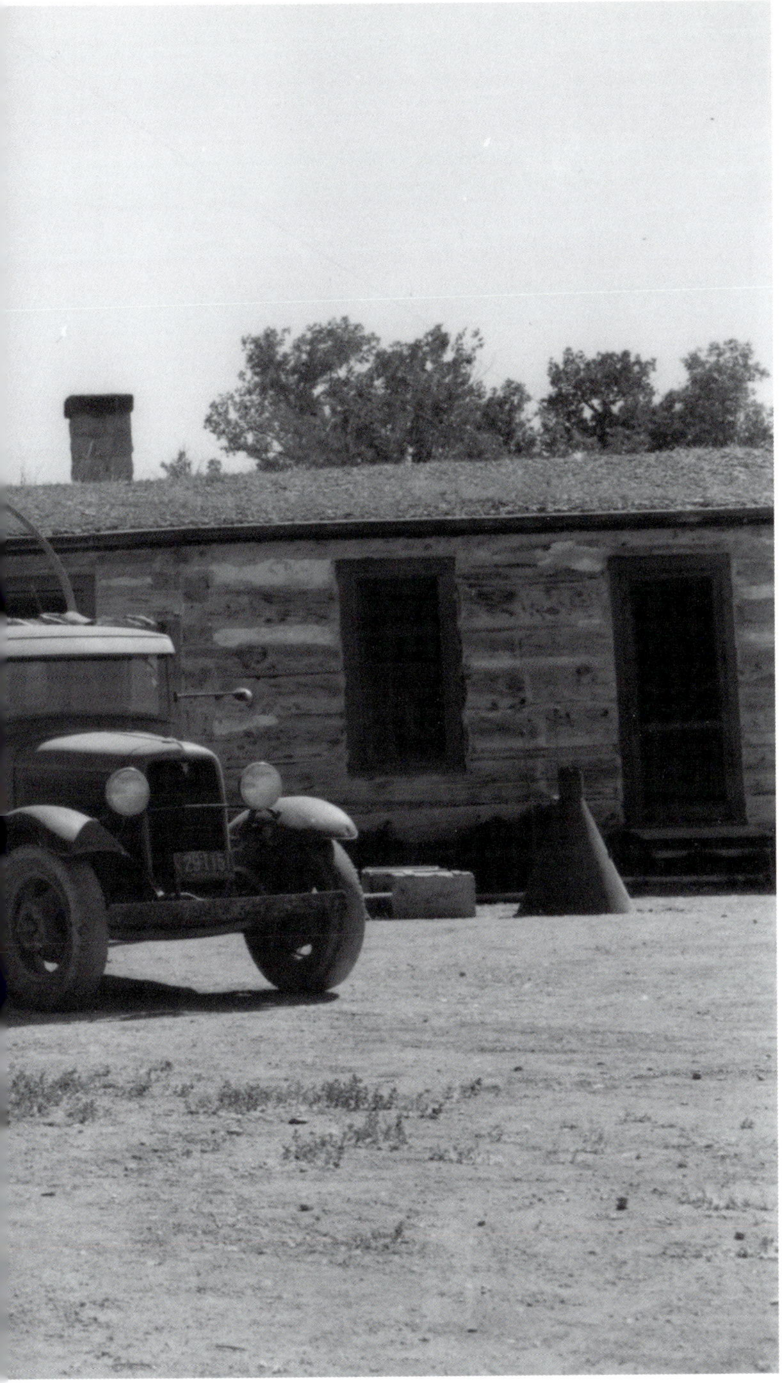

Loading the roundup truck. Quarter Circle U Ranch, Montana. Arthur Rothstein, June 1939.
LC-USF34-027591

Cook and hired girl. Quarter Circle U Ranch, Montana. Arthur Rothstein, June 1939.
LC-USF34-027470-D

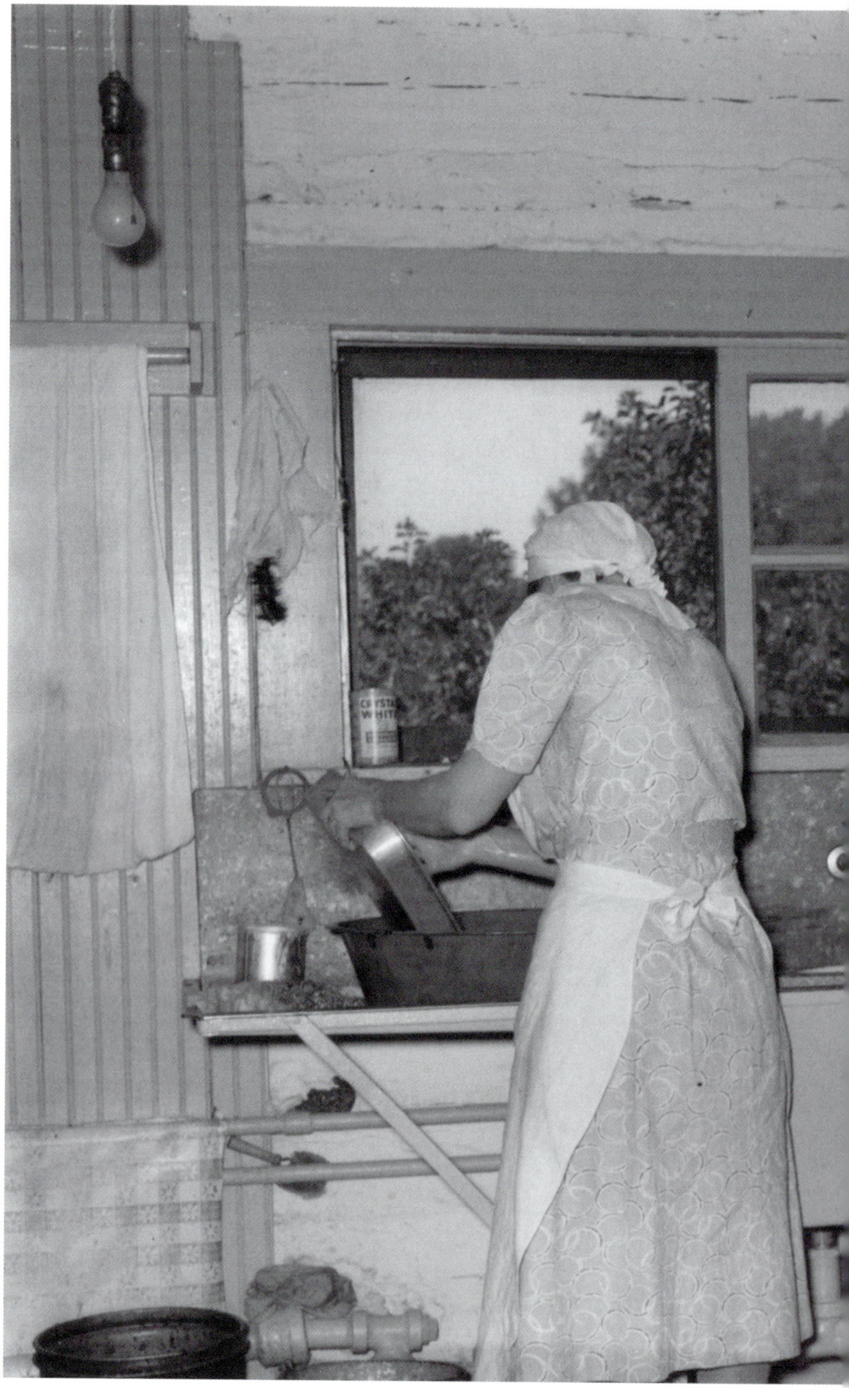

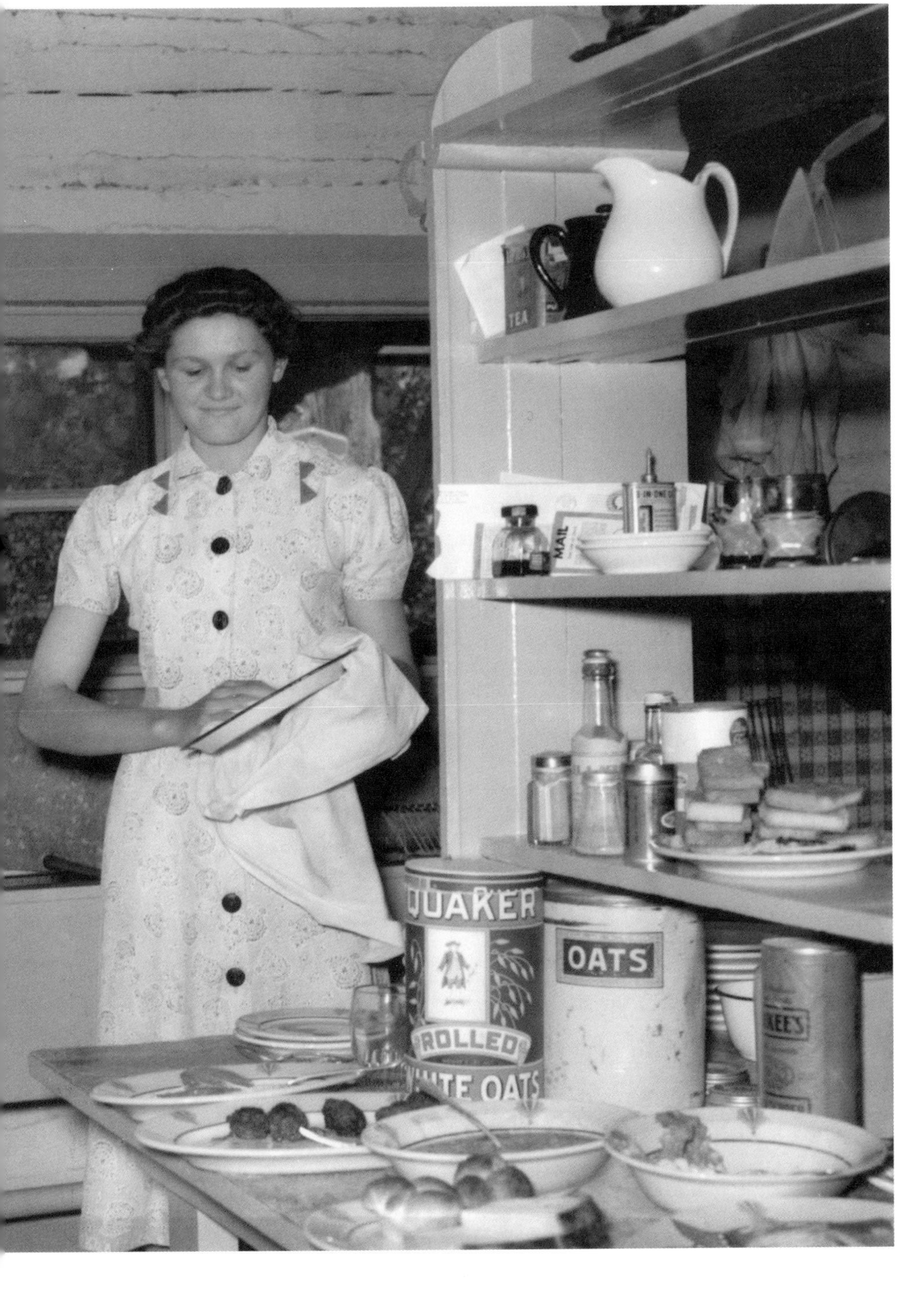
TEA
MAIL
QUAKER
ROLLED
WHITE OATS
OATS

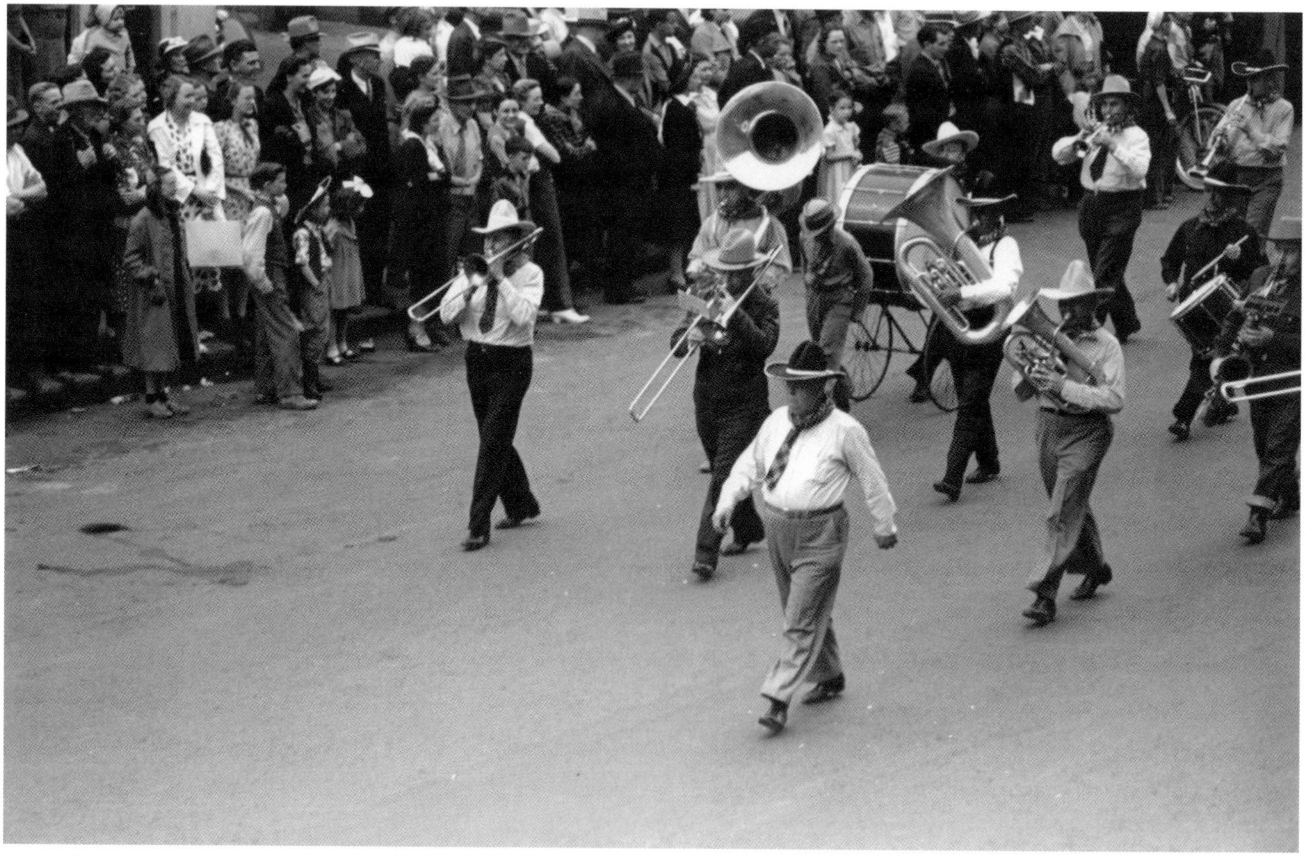

ABOVE: Cowboy band in Go Western parade. Billings, Montana. Arthur Rothstein, Summer 1939.
LC-USF33-003074-M4 DLC

OPPOSITE: Farmer with tractor. Fairfield Bench Farms, Montana. Arthur Rothstein, May 1939.
LC-USF34-027287-D DLC

THE TRACTOR

The tractor fundamentally changed the nature of farm work and, to the extent that ranchers raised hay, much ranch work as well. The first big open-geared gas tractors came into use in areas of extensive farming between 1910 and 1915. All-purpose, rubber-tired tractors with complementary machinery came into wide use in the 1930s. While horses and mules provided power for agricultural work, they ate up more than twenty percent of the crops they helped grow. It has been estimated that a tractor would replace about five horses or mules during the early 1900s. In the 1930s, farmers and ranchers were strapped for cash by the Depression, just like the rest of the nation. In the 1940s, as the economy began to improve and more ranchers and farmers could afford them, tractor sales increased, and the careers of most draft animals ended. The number of horses peaked at just over 25 million animals around 1920. About that same time, the number of tractors began rising and peaked at just under 5 million in the late 1960s and 1970s. The amount of tractor power overtook the amount of horsepower on American farms and ranches in 1945.

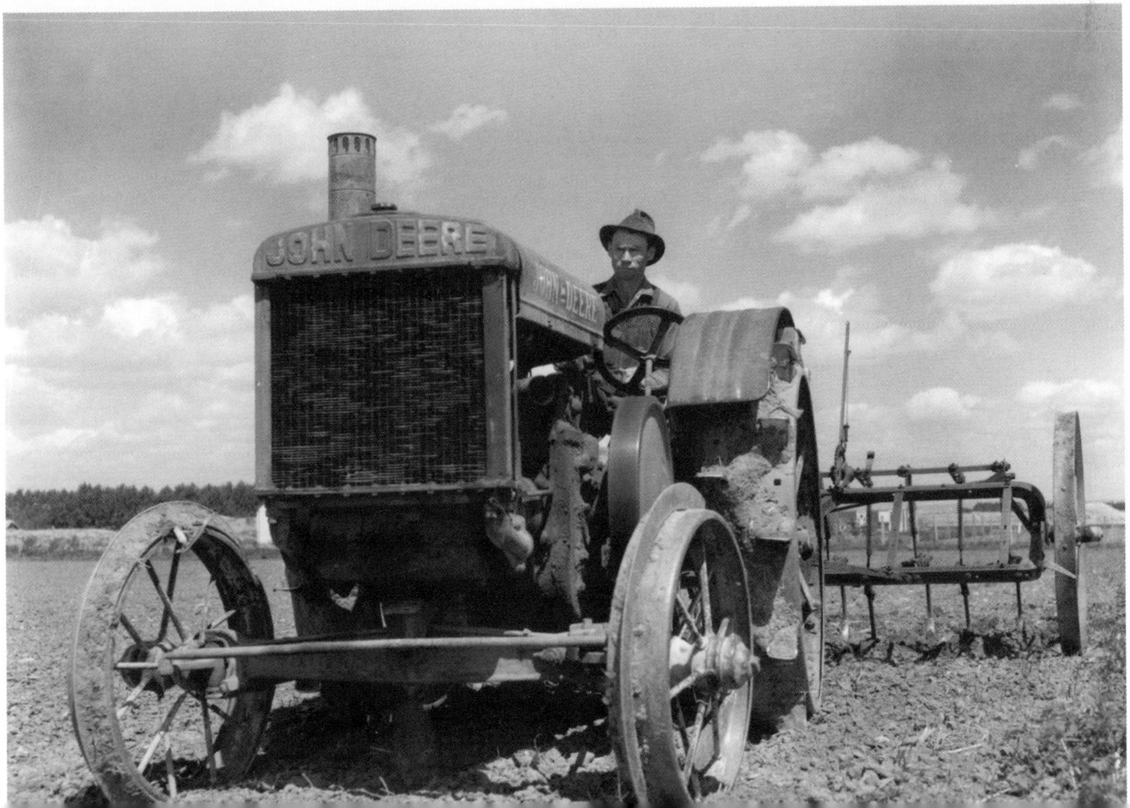

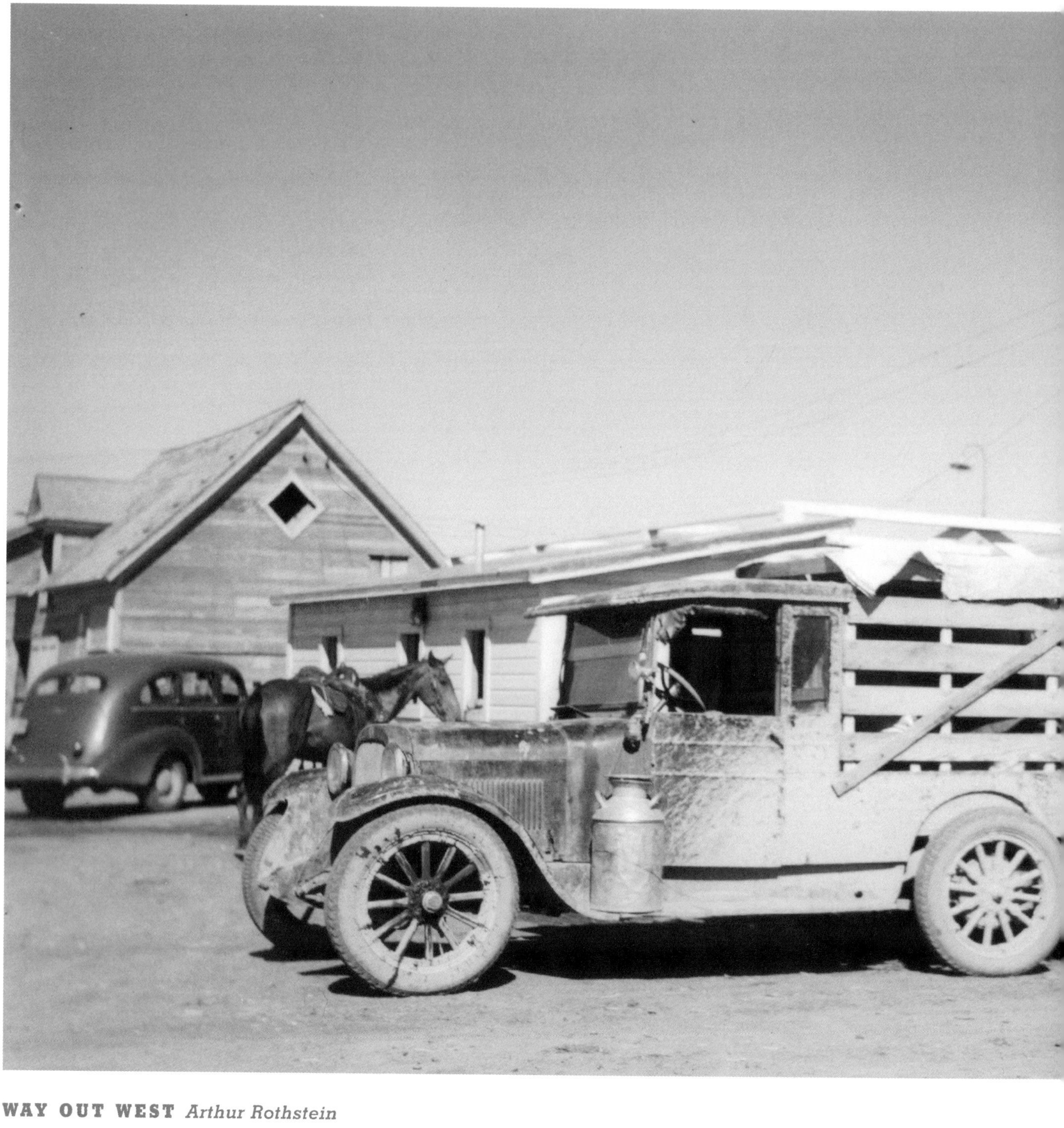

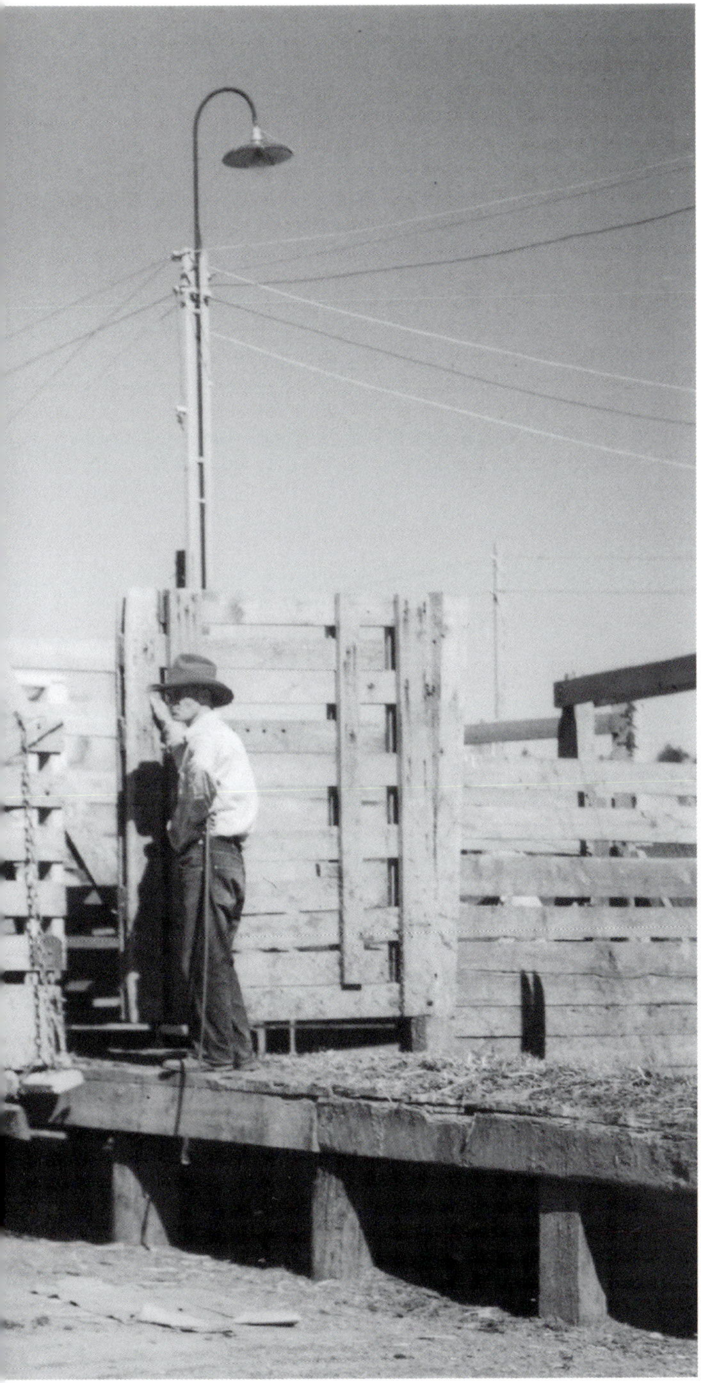

Trucks like these bring the last remnants of the small ranchers' herds to the stockyards. Billings, Montana. Arthur Rothstein, July 1936.
LC-USF34-005017-D DLC

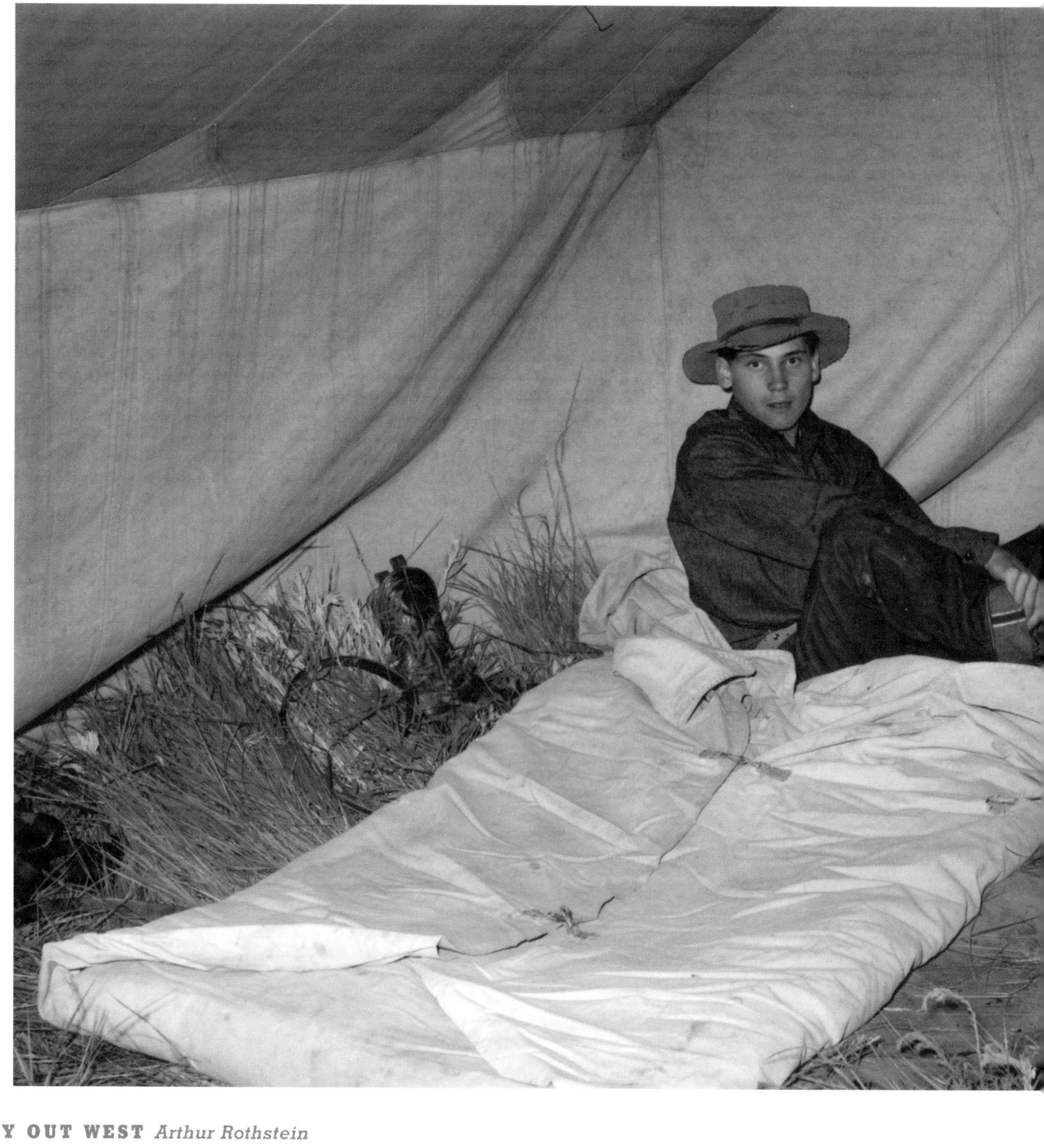

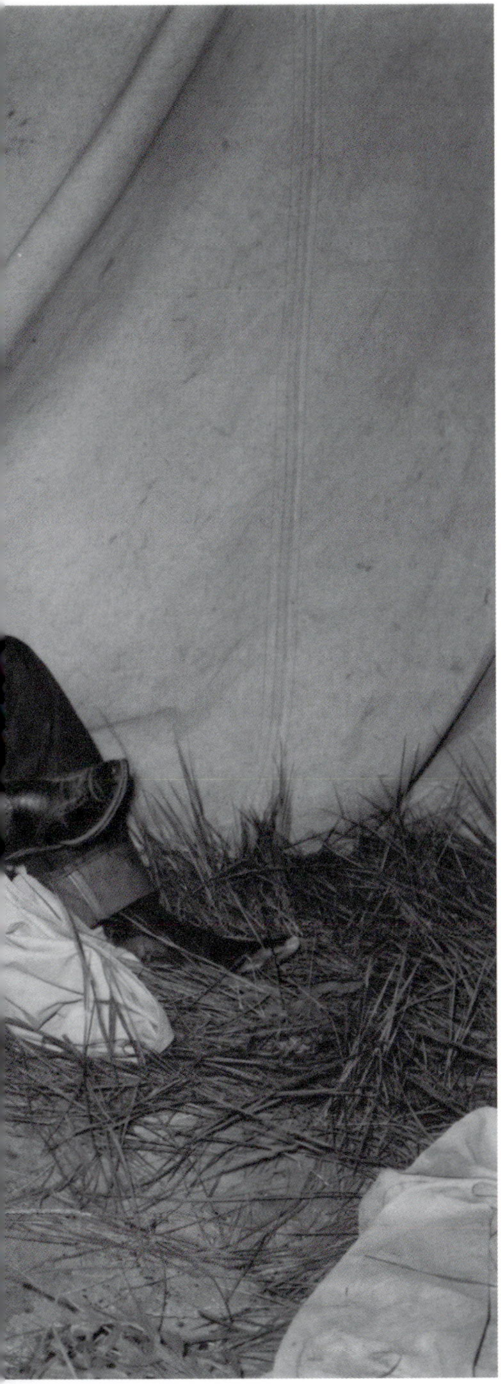

Cowboy going to bed. Quarter Circle U Ranch roundup. Big Horn County, Montana. Arthur Rothstein, June 1939.
LC-USF34-027841-D

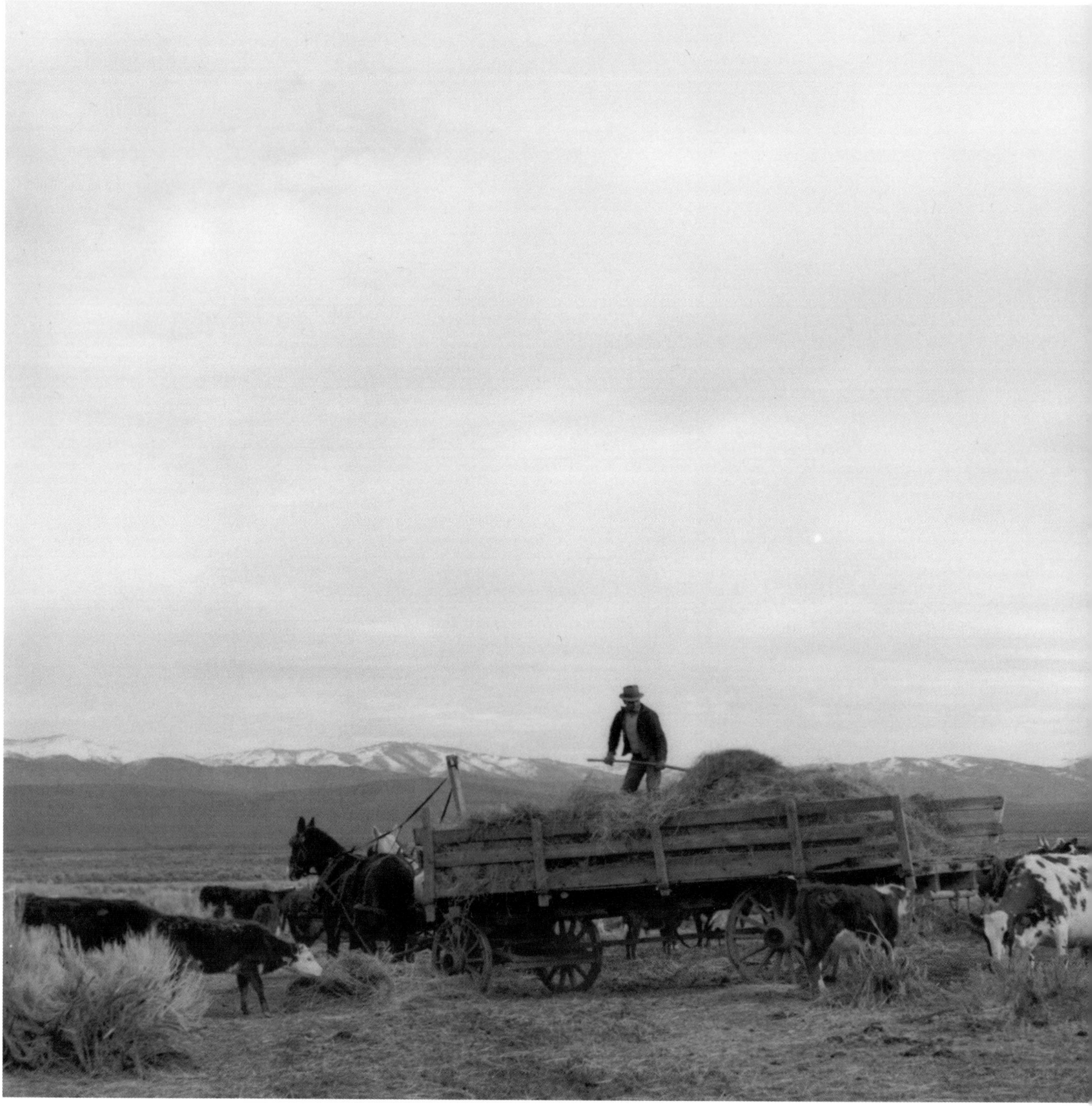

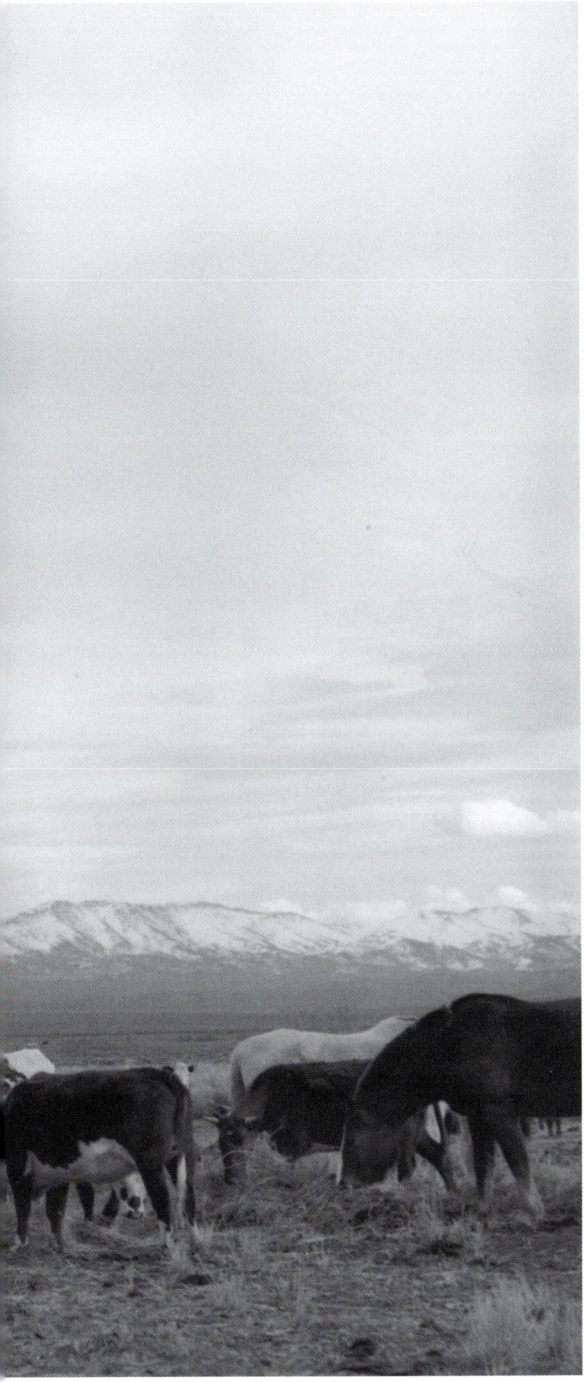

Feeding cattle. Douglas County, Nevada. Arthur Rothstein, March 1940.
LC-USF34-029709-D

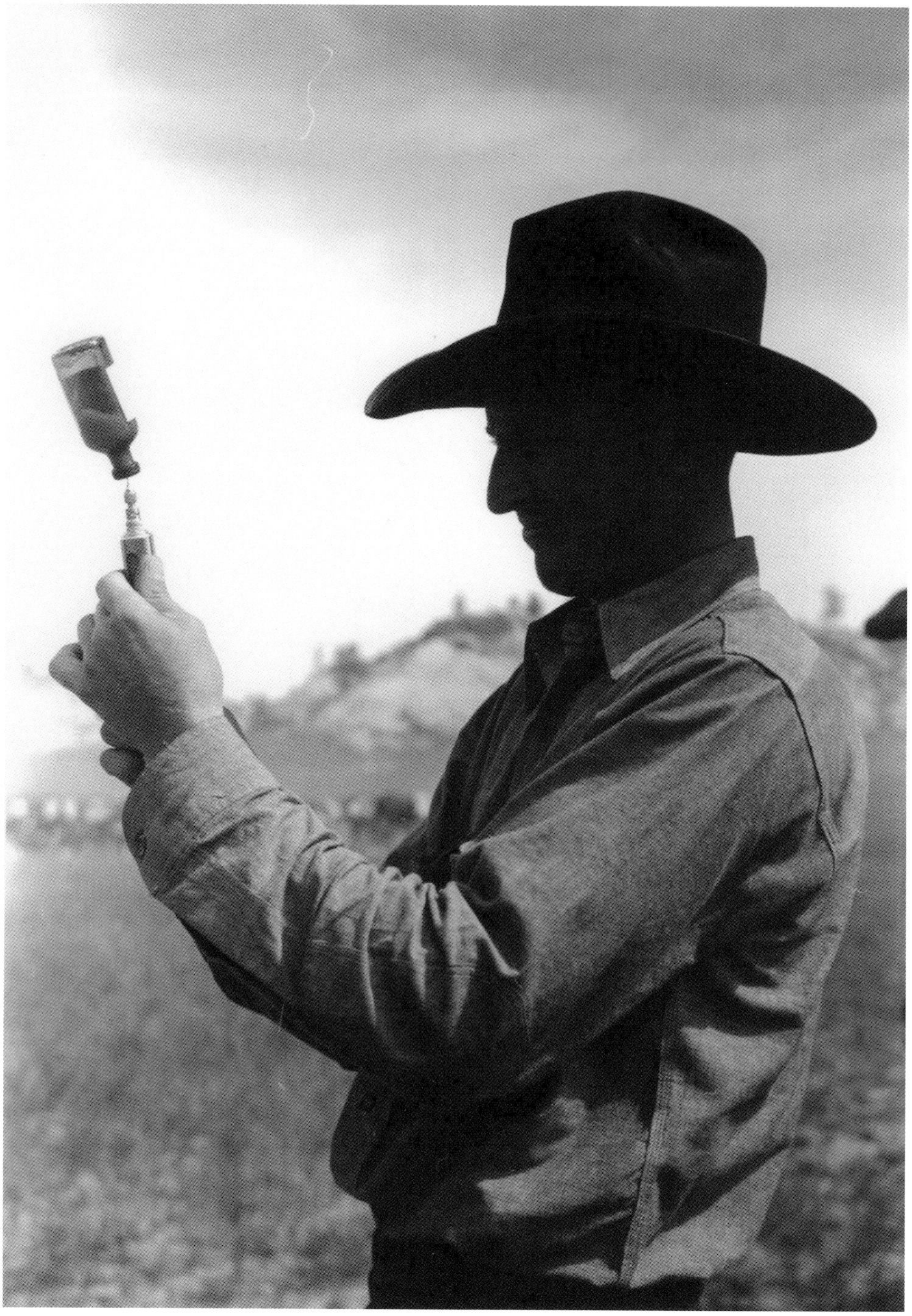

 Arthur Rothstein

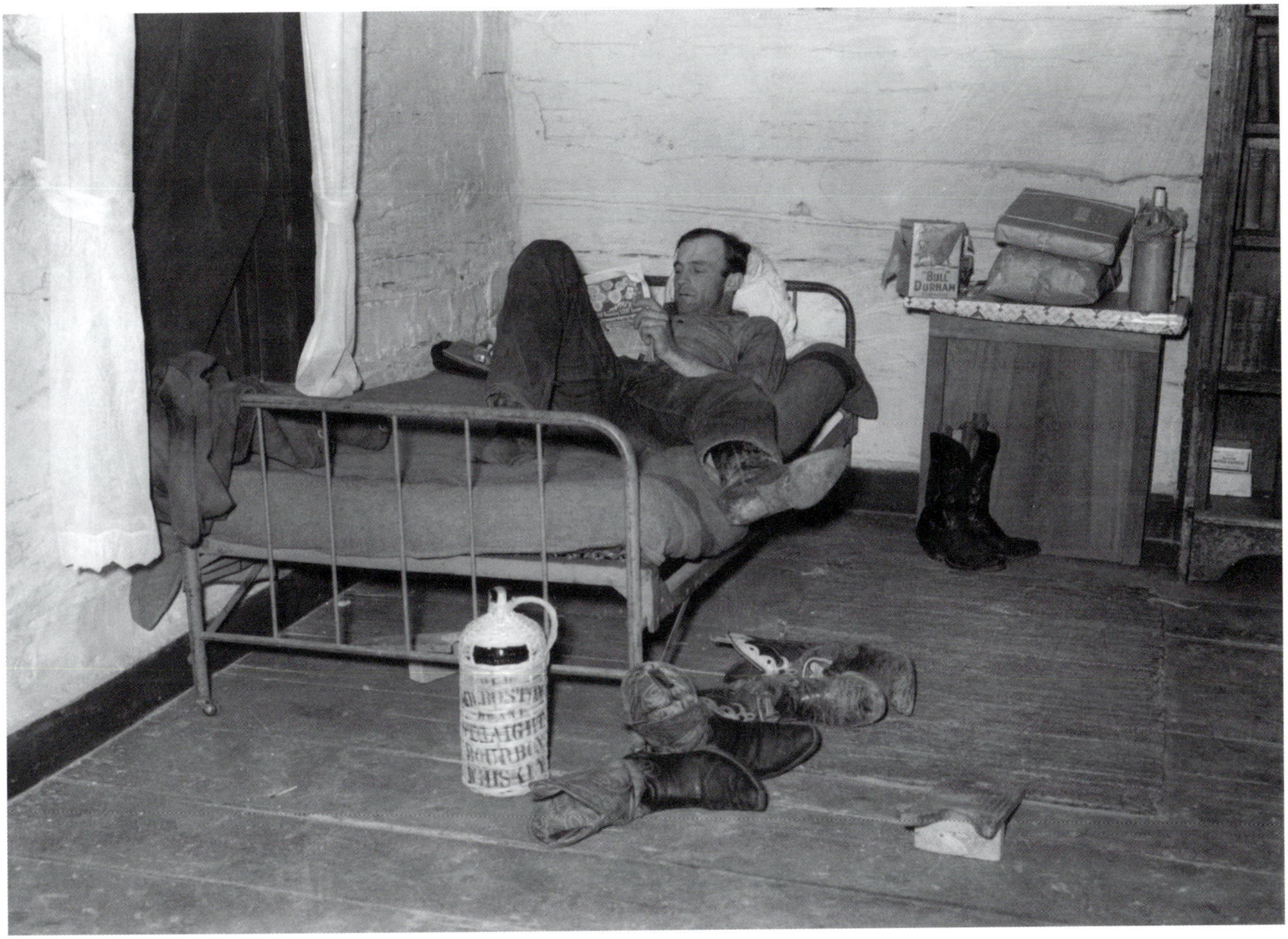

OPPOSITE: Cowboy filling syringe with black leg serum. Quarter Circle U Ranch roundup, Montana. Arthur Rothstein, June 1939.
LC-USF33-003241-M1

ABOVE: Cowhand at ease in bunkhouse. Quarter Circle U Brewster-Arnold Ranch Company, near Birney, Montana. Arthur Rothstein, June 1939.
LC-USF346-027447-D

RESOURCES AND REFERENCES

Artist Biographies

John Collier Jr.

http://nomiedo.blogspot.com/2014/06/john-collier-jr-photographer.html (accessed June 3, 2016).

https://en.wikipedia.org/wiki/John_Collier_Jr. (accessed June 3, 2016).

https://www.flickr.com/people/johncollierjr/ (accessed June 3, 2016).

Russell Lee

http://www.cimarronkansas.net/russelllee.htm (accessed May 26, 2016).

John Vachon

International Center of Photography, https://www.icp.org/browse/archive/constituents/john-vachon (accessed June 3, 2016).

http://www.mnopedia.org/person/vachon-john-1914-1975 (accessed June 3, 2016).

Marion Post Wolcott

http://people.virginia.edu/~ds8s/mpw/mpw-bio.html (accessed June 3, 2016).

https://en.wikipedia.org/wiki/Marion_Post_Wolcott (accessed June 3, 2016).

Websites

Farm Security Administration/Office of War Information. http://www.loc.gov/pictures/collection/fsa/docchap2.html (accessed June 3, 2016).

Indiana University Bloomington. http://www.iub.edu/~iuam/online_modules/fsa/fsaartists.html (accessed June 4, 2016).

Museum of Contemporary Photography. http://www.mocp.org/detail.php?t=objects&type=group&f=&s=&record (accessed May 26, 2016).

New Mexico Tells New Mexico History. http://online.nmartmuseum.org/nmhistory/art/sorting-beans-juan-lopez-and-son-tra (accessed June 3, 2016).

Books

Brown, Mark H., and W.R. Felton. *Before Barbed Wire: L.A. Huffman, Photographer on Horseback.* New York: Henry Holt and Co., 1956.

Collier, John, Jr. *Visual Anthropology: Photography as a Research Method.* New York: Holt, Rinehart and Winston, 1967.

Curtis, James. *Mind's Eye, Mind's Truth: FSA Photography Reconsidered.* Philadelphia: Temple University Press, 1989.

Egan, Timothy. *The Worst Hard Time: The Untold Story of Those Who Survived the Great American Dustbowl.* Boston and New York: Houghton Mifflin, 2006.

Fischer, Claude S. *America Calling: A Social History of the Telephone to 1940.* University of California Press, 1992.

Fleischhauer, Carl, and Beverly W. Brannan, eds. *Documenting America, 1935–1943.* Berkeley: University Press of California in association with the Library of Congress, 1988.

Hudson, Lois Phillips. *Reapers of the Dust: A Prairie Chronicle.* Little Brown & Co., 1965.

Hurley, F. Jack. *Portrait of a Decade: Roy Stryker and the Development of Documentary Photography in the Thirties.* Baton Rouge: Louisiana State University Press, 1972.

Kyvig, David E. *Daily Life in the United States, 1920–1940: How Americans Lived Through the Roaring Twenties and the Great Depression.* Chicago: Ivan R. Dee, 2002.

Mora, Gilles, and Beverly W. Brannan. *FSA: The American Vision.* New York: Abrams, 2006.

Nye, David E. *Electrifying America: Social Meanings of a New Technology 1880–1940.* Cambridge: MIT Press, 1990.

Price, B. Byron. *Erwin E. Smith, Cowboy Photographer.* Fort Worth: Amon Carter Museum, 1992.

Steichen, Edward. *The Family of Man.* New York: The Museum of Modern Art, 1955.

Stryker, Roy E., and Nancy C. Wood. *In This Proud Land: America, 1935–1943, as Seen in the FSA Photographs.* New York: Galahad Books, 1973.

ABOUT THE AUTHOR

Charlie Seemann is the Executive Director Emeritus of the Western Folklife Center in Elko, Nevada. He holds an M.A. in folklife studies from the University of California at Los Angeles. He was an instructor in folklore and folk music at Moorpark College in California for six years and then worked as a folklorist and Director of the Western Regional Folklife Festival for the National Park Service/Golden Gate National Recreation Area in San Francisco until 1981. He then moved to Nashville, Tennessee, where he was Deputy Director at the Country Music Foundation/Country Music Hall of Fame for twelve years.